P

Bio-Glasses

An Introduction

Edited by

Julian R. Jones
Department of Materials, Imperial College London,
London, UK

Alexis G. Clare
New York State College of Ceramics, Alfred University,
Alfred, New York, USA

A John Wiley & Sons, Ltd., Publication

This edition first published 2012
©2012 John Wiley and Sons, Ltd.

Registered office
John Wiley & Sons, Ltd., The Atrium, Southern Gate, Chichester, West Sussex, PO19 8SQ, United Kingdom

For details of our global editorial offices, for customer services and for information about how to apply for permission to reuse the copyright material in this book please see our website at www.wiley.com.

Library of Congress Cataloging-in-Publication Data

Bio-glasses : an introduction / [edited by] Julian Jones and Alexis G. Clare.
    p. ; cm.
Includes bibliographical references and index.
ISBN 978-0-470-71161-3 (cloth)
I. Jones, Julian R. II. Clare, Alexis.
[DNLM: 1. Biocompatible Materials. 2. Glass. 3. Drug Delivery Systems. 4. Regeneration.
5. Tissue Scaffolds. QT 37]
    610.28–dc23

                2012008850

A catalogue record for this book is available from the British Library.

ISBN: 978-0-470-71161-3

Typeset in 10.5/13 Sabon by Laserwords Private Limited, Chennai, India
Printed and bound in Singapore by Markono Print Media Pte Ltd

Contents

List of Contributors

Aldo R. Boccaccini
Department of Materials Science and Engineering, University of Erlangen-Nuremberg, Erlangen, Germany

Delia S. Brauer
Otto Schott Institute, Friedrich Schiller University Jena, Jena, Germany

Qi-Zhi Chen
Department of Materials Engineering, Monash University, Clayton, Victoria, Australia

Alexis G. Clare
Kazuo Inamori School of Engineering, New York State College of Ceramics, Alfred University, Alfred, USA

Alastair N. Cormack
Kazuo Inamori School of Engineering, New York State College of Ceramics, Alfred University, Alfred, USA

Delbert E. Day
Center for Bone and Tissue Repair, Graduate Center for Materials Research, Materials Science and Engineering Department, Missouri University of Science and Technology, Rolla, Missouri, USA

Satoshi Hayakawa
Department of Bioscience and Biotechnology, Okayama University, Tsushima, Okayama, Japan

Wolfram Höland
Ivoclar Vivadent AG, Schaan, Principality of Liechtenstein

Leena Hupa
Process Chemistry Centre, Åbo Akademi University, Biskopsgatan, Åbo, Finland

Julian R. Jones
Department of Materials, Imperial College London, South Kensington Campus, London, UK

Steven B. Jung
MO-SCI Corporation, HyPoint North, Rolla, Missouri, USA

Matthew D. O'Donnell
Department of Materials, Imperial College London, South Kensington Campus, London, UK

Akiyoshi Osaka
Department of Bioscience and Biotechnology, Okayama University, Tsushima, Okayama, Japan

Yuki Shirosaki
Department of Bioscience and Biotechnology, Okayama University, Tsushima, Okayama, Japan

Kanji Tsuru
Department of Biomaterials, Kyushu University, Maidashi, Higashi, Fukuoka, Japan

Enrica Verné
Department of Applied Science and Technology, Politecnico di Torino, Torino, Italy

Antti Yli-Urpo
Institute of Dentistry, Faculty of Medicine, University of Turku, Turku, Finland

Foreword

'Hamburger. Hot Dog. Ice Cream.' Five ordinary words, but each had extraordinary significance. It was 1984 when Dr. Gerry Merwin, MD, an Ear, Nose and Throat surgeon at the University of Florida, Gainesville, Florida, whispered the words into the ear of a patient. She was an extraordinary patient – a young mother, expectant with her second child. She was desperate to be able to hear her new-born baby cry. However, the mother was deaf from an infection that had dissolved two of the three bones of her middle ear. Only part of the stapes (stirrup) remained. Under a local anaesthetic, Dr. Merwin had just implanted the world's first Bioglass® device into her middle ear. The implant was designed to conduct sound waves from her eardrum to her inner ear, the cochlea, and thus restore her hearing.

Bioglass was the first man-made material to bond to living tissues. I discovered it in 1969, but the material had to pass 15 years of *in vitro* (cells growing on it in the laboratory) and *in vivo* (animal) tests before the first device could be implanted. The University of Florida Shands Hospital Ethics Committee had approved the first human trial after evaluating data from safety tests done on hundreds of mice by Dr. Merwin, his surgical residents and Dr. June Wilson, who interpreted the histology data.

But, the big questions remained: 'Would this new material work in a human? Would the implant bond to the soft connective tissue of the eardrum? Would it bond to the hard connective tissue (bone) of the stapes? Would it conduct sound? Would normal hearing be restored?'

No one knew the answers to these questions. Ear surgeons had inserted other types of middle ear implants in patients for many years, but they often failed. The materials used were metals and plastics, selected because they were as inert and non-toxic as possible in the body. A thin layer of scar tissue formed around the metal and plastic parts, isolating them from the body, eventually causing the implant to be forced out of position. All of the first-generation biomedical materials used in the body (so-called bio-inert materials) led to the formation of scar tissue. For some clinical needs, the scar tissue poses no problem. For a middle ear implant, scar tissue can be disastrous. Continual vibration and motion of the implant can wear a hole in the eardrum. The implant can come out through the hole, permanently damaging the eardrum.

The Bioglass middle ear implant tested a new concept in repair of the human body, *bioactive bonding*. The special composition of the glass contained the same compounds as present in bones and tissue fluids: Na_2O, P_2O_5, CaO and SiO_2. When a bone is broken, the body uses these compounds to form new bone. The new Bioglass implant released these compounds and the cells in the native bone used them to form a bioactive bond. The collagen of soft connective tissues, such as the tympanic membrane, also bonded to the bioactive hydroxyapatite layer that forms on the bioactive glass surface.

The theory underlying a second generation of biomedical materials, based upon bioactive bonding, was ready for its final test.

Dr. Merwin whispered the five words. A big smile appeared on the face of the patient and she repeated: 'Hamburger. Hot Dog. Ice Cream!' The Bioglass middle ear implant worked.

Ten years later in a follow-up study, the implant was still working, and the mother could hear her 10-year-old child laughing and singing. In the years since, thousands of patients have had their hearing restored with bioactive middle ear implants. The field of medicine and the nature of biomaterials had been changed forever.

It is now nearly 30 years since this first, epochal human trial. The speciality field of bioactive materials has expanded exponentially in those years. Millions of patients have undergone various types of repair and reconstructive surgery using formulations of bioactive materials, such as Bioglass, synthetic hydroxyapatite, Si-substituted hydroxyapatite (Actifuse®), tricalcium phosphates (e.g. Vitoss®), bioactive glass–ceramics (A/W glass–ceramic, Cerabone®), and so on. However, the grandfather material, 45S5 Bioglass, is still the material with the highest level of bioactivity and the fastest rate of bioactive bonding, and for some applications is the so-called 'gold standard'. Bioglass is now used as a synthetic bone graft (e.g. NovaBone®) and it can now be found as the active ingredient (NovaMin®) in a market leading brand of toothpaste for sensitive teeth. The soluble glass dissolves and seals the tubules in exposed dentine, preventing exposure of nerve endings to hot and cold food and drinks.

Thus, understanding the science, technology and applications of bioactive glasses is a very important educational need for the healthcare and glass community. Many new developments have occurred during the past 30 years that are not discussed in standard materials science textbooks. Many subjects, such as sol–gel processing of bioactive gel–glasses, genetic stimulation of osteogenesis by ionic dissolution products of bioactive glasses, stimulation of angiogenesis, bioactive composites, hybrid bioactive materials, phosphate glasses, bioactive materials with hierarchical porosity, molecular modelling of glass structures and bioactivity mechanisms, tissue engineering and regenerative medicine, are now important topics in the field that did not even exist as concepts in 1984. Also, other important biomedical glass and glass–ceramic systems for therapeutic treatment of tumours and repair of diseased and damaged teeth are in widespread use and enhancing the quality of life for millions of patients throughout the world.

This important new book provides a basic level of understanding of all of the above topics. Of special importance is the fact that this book assumes that the reader is just getting started in the field. It is a primer. It provides the necessary foundation of science and technology at a beginning level in order for the reader to explore later the multitude of papers being published annually in this new field. Without a basic understanding, such as provided by this book, a person is easily confused. The reason is that the interface generated over time between a bioactive glass and the body is controlled by an integrated synthesis of inorganic chemistry, physical chemistry and biochemistry. The man-made material and the living material become as one at a molecular level. This type of 'living interface' mimics that between hard and soft connective tissues in the body that has evolved over billions of years. This unique character of bioactive bonding requires a unique textbook in order to comprehend and explore these materials and their clinical use. This unique book provides the fundamental level of comprehension needed. I hope it encourages the bright young creative minds of the future to enter the field and take bioactive glasses and related materials forward to the next generation of medical devices and continue to improve the quality of life of patients. For the experienced researcher, the book provides a comprehensive overview of the important current topics in the field written by world-class authors. Unlike many conference proceedings, this volume has been written by carefully selected contributors who have created much of the subject matter they discuss in their chapters, and as a consequence the contents are authoritative.

To all readers, beginner or experienced: read, enjoy and marvel at this wonderful material!

Larry Hench
Inventor of Bioglass®
20 September 2011

Preface

I found out about 'Bioglass' by accident. I was giving a presentation in a lecture competition in London, while I was an undergraduate at Oxford. After my talk – the subject of which (spray forming of aluminium alloys) is incidental – I began chatting to one of my fellow contestants about his biomaterials presentation. As I expressed interest in the work, Larry Hench overheard and began telling me about his own work. I was captivated and decided I should do all I could to do a PhD in Larry's group. It was a while later (a few weeks actually, due to Larry's humble nature) that I realised that he invented Bioglass and was a founder of the field of 'bioactive ceramics'. Even though I was studying materials science in a top university, I had not heard of Bioglass or the many variants that had been developed since its invention. It should not be left to chance events for young people to come across these important and exciting materials.

So, one aim for us in writing this book is to make more people aware of bio-glasses and their variants, their application in medicine and their great potential for future clinical procedures. The book covers a wide range of material, from what a glass is, through the origins of bioactivity and how bioactive glasses can regenerate bone and heal wounds, to glasses used in cancer treatment and new-generation materials for dental reconstruction and tissue engineering. Other books available, at the time of writing, either seemed to try to cover too broad an area – such as all bioceramics – or were a collection of articles collated at conferences on the very latest developments in the field, which were perhaps not accessible to the non-expert.

We set out to produce a book that was accessible to those curious about materials in medicine, whether their background was scientific, engineering or medical. We hope that undergraduate students will find the book interesting, and decide that this is an area about which they would like to discover more or in which perhaps they would like to follow a career. The international profile of bio-glasses is increasing all the time, so more and more healthcare professionals will be exposed to bio-glass-related treatments and devices. Owing to its introductory and accessible nature, this book will be a useful tool for healthcare professionals to quickly learn about bio-glasses and their potential. Anyone brushing their teeth with the latest generation of toothpastes (containing fine Bioglass particles) may also be curious to discover how the active ingredient works in treating sensitive teeth.

I would like to take this opportunity to thank a few people. First, my co-editor, Alexis Clare – who came up with the concept for this book – as it would not have been possible without her. Alexis and I are both very pleased that Larry was so willing to write the Foreword to this book in his 'retirement', while he works on new projects developing materials for soaking up and recycling oil slicks, materials for supercapacitors and, of course, writing his *Boing Boing, the Bionic Cat* children's stories that introduce science concepts to children. We would also like to thank the other members of Technical Committee 4 (TC04) of the International Commission of Glass (ICG) for their contributions, many of them writing chapters for this book, and of course the ICG itself.

Julian R. Jones
Senior Lecturer, Department of Materials, Imperial College London
and Visiting Professor at Nagoya Institute of Technology, Japan
Chair of Technical Committee 4 (TC04)
of the International Commission of Glass (ICG)
March, 2012

1

The Unique Nature of Glass

Alexis G. Clare
Kazuo Inamori School of Engineering, New York State College of Ceramics, Alfred University, Alfred, USA

1.1 WHAT IS GLASS?

We tend to think of glass as a single material from which we manufacture many useful articles, such as windows, drinking vessels, and storage containers that can contain quite corrosive liquids, including aggressive laboratory chemicals. Therefore, the glass has to be quite corrosion-resistant and inert, including being able to maintain its optical properties while being in aggressive environments such as a dishwasher or extreme weather. For a material that we generally view as "delicate," in terms of attack from chemicals, it can be quite resistant. Another extreme environment is the human body. An implanted medical device is subjected to a warm and wet environment with continual fluid flow and complex mechanical loads, but perhaps more importantly there are many cells, some at work to reject foreign inert materials. They do this by encapsulating the materials with fibrous (scar) tissue. Hip replacements generally last 15 years or so, and in 2009 there was the story of a man who cut himself shaving and out of his chin fell a piece of glass. He had a lump under his chin, but he thought it was an abscess. In fact, 20 years

Bio-Glasses: An Introduction, First Edition. Edited by Julian R. Jones and Alexis G. Clare.
© 2012 John Wiley & Sons, Ltd. Published 2012 by John Wiley & Sons, Ltd.

earlier he was involved in a car accident and a piece of the windscreen embedded in his chin, unbeknown to him. The inert glass would have been sealed off from the body by fibrous tissue and over the years it was pushed out through the soft tissue to the skin.

This book will illustrate that, for biomedical applications, certain glasses can be active in the body and stimulate the natural healing of damaged tissues. The degradation of the glass is actually encouraged and plays an important role, allowing space for tissue regrowth and actively stimulating cells to produce tissue.

Glass is actually far from being a single composition but rather is a state of matter, a subset of the solid state. A glass is a network of atoms (most commonly silicon) bonded to each other through covalent bonds with oxygen atoms. A silica-based glass is formed of silica tetrahedra (Figure 1.1) bonded together in a random arrangement. Window glass is usually based on the soda–lime–silica (Na_2O–CaO–SiO_2) system. Bioactive glasses also often contains these components, but in different proportions to inert glasses. Chapter 5 discusses the atomic structure in more detail.

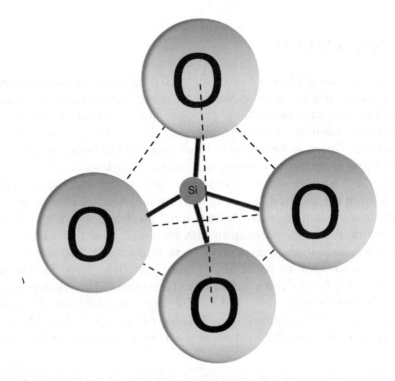

Figure 1.1   Schematic of a silica tetrahedron, the basic component of a silica glass.

Glass differs from what we think of as regular (crystalline) solids in a number of ways. A glass does not "melt" in the way a crystalline solid does. If we heat a pure single-phase crystalline solid, at some point the solid will melt with a well-defined melting temperature. Impurities will usually alter the melting point, and the presence of more than one crystal phase will lead to multiple melting points. Nevertheless, when melting occurs, there is an abrupt change from the solid to the liquid. If we attempt the same experiment with a piece of glass, we will not see a sudden change at a well-defined temperature, but we will see the solid "ease" into the liquid, probably a quite viscous liquid. The glass "transition" from the solid glass to the viscous liquid glass is an important property. Basically, the glass is an elastic solid below this transformation region and a viscous liquid above it. The structure of the solid has all the attributes of a liquid, except that the solid does not have the ability to flow on any meaningful time scale. The apocryphal story of cathedral windows in Europe being thicker at the bottom than they are at the top, having flowed due to gravity, is not true: the silicate glasses in windows are only going to flow on something approaching a geological time scale (unless things were really to heat up on Earth, in which case we would not be worrying about cathedral windows). What is even more curious about this glass transition (called T_g for short) is that, unlike a melting point, the range over which it happens and the temperature at which it starts very much depends on how the glass was made in the first place, the rate at which the glass was cooled, whether it had subsequent heat treatments, and so on. For most commercial glass used in medicine and biotechnology, if the glass is cooled from the melt faster, the overall glass structure will have a larger volume (lower density) than one that is cooled slowly.

In terms of structure, solid glass and liquid glass look very similar. However, if you were able to take a photograph of the atoms showing their position, in a subsequent photograph of a liquid the atoms would have all moved, whereas in the glass they would be in much the same position as in the first photograph. Essentially, a glass is an elastic solid without the structural periodicity and long-range order of a crystalline material. It looks like a liquid but behaves like a solid.

Why are not all solids like this? After all, it seems that there is a lot less rearranging involved in moving through the glass transition than there is in melting a crystalline material. Thermodynamics provides the clue: thermodynamically, systems are generally driven to the lowest energy (stable) state, so most solids would adopt the inherent order of the crystalline state, resulting in a lower potential energy for the solid.

However, kinetics occasionally gets the chance to overrule thermody-namics and will not allow the ordered structure to form if there is not enough time to arrange the atomic structure and establish the ordered state: hence the American Society for Testing and Materials (ASTM) definition of a glass as being a material that has "cooled from the melt without crystallizing." The logical question would then be: "How fast would one have to cool for kinetics to overcome thermodynamics?" The speed of this would be dependent upon the composition. Silicate melts can cool relatively slowly without crystallizing (about 20 degrees per minute), whereas for a bulk metallic glass it is more like 2 degrees per second. So, if one has to thermodynamically trick a material into being a glass, what are the advantages?

The word "glass" evolves from the Latin word *glacies* meaning "ice," and by far the most utilized property of glass is its transparency, which comes as a result of its inherent isotropic nature. Although the atoms are not organized and are generally quite randomly arranged, the glass as a whole has a similar structure throughout. While glass can in principle be made from any mixture of atoms, the majority of commercial glasses are based upon silicates, and these have a transparency from just into the ultraviolet to somewhere in the infrared, with a transmission typically of up to 90%. The clarity of some types of glass used to make optical fibers is such that the fiber is transparent for miles and miles. A piece of window glass does not have 100% transparency in the visible mostly due to reflection loss. The reflection of light from glass depends upon the refractive index, which is the ratio of the speed the light moves through a vacuum compared to its speed in the material. Glasses with higher refractive index reflect more light, and this property is often used for aesthetic reasons. For example, the "lead crystal" that is often used in fine wine glasses is very sparkly due to the high refractive index of the lead-containing glass. The name "lead crystal" is a little deceiving, as glasses are certainly not crystalline – they are amorphous in structure. Reflection loss can be cut down by adding an anti-reflection coating, which is a coating that is based upon the destructive interference of light waves reflected from two interfaces. The limits of transparency in the ultraviolet and the infrared are governed by the electrically insulating natures of the glass and the type of elements and their bonds, respectively. Typically, the more electrically insulating a glass, the better ultraviolet transparency it exhibits. The heavier the elements and the lower the force constant of the bonds in the glass, the more infrared transparent the glass is. Another detriment to transparency is the existence of coloring ions. These are typically either transition metals or rare earths, the transition metals being very strong coloring agents. Hence, for applications where thick optical paths are needed, then highly pure glass is required, because there are common contaminants such as iron in the silica sand used to make glass.

The second most utilized property of glass is outstanding chemical durability in a number of different chemical environments. However, it should be noted that the chemical durability of glass is not always quite as outstanding as is sometimes believed. As will be discussed later on, glass does not always either require or desire high chemical durability. However, in comparison to many other materials, the chemical durability of glass is very good, and its isotropic nature lends itself to both the high durability and the control of the lower durability.

One of the lesser extolled attributes of glass is its ability to be engineered to meet need. Unlike a crystalline material, in which phases tend to have very well-defined and rigorously maintained stoichiometry (e.g. $K_2O \cdot 2SiO_2$). That is not to say that one can make a SiO_3-based glass, but that potassium silicate glass can be made over a continuous range of potassia to silica ratios with properties that will also exhibit a fairly continuous range of values between those of the end members. Despite the ubiquitous presence of glass in our daily lives, outside of the glass community, few people are aware of the staggering array of compositions there are for glass, from plain old silica to exotic glass formulations containing many components that might not even contain oxygen. Even more astounding are the advances that have been made in metallic glasses over recent years. There is an ever-increasing number of metallic alloys that can be made in bulk form, with an amorphous structure that results in extremely interesting mechanical properties and, in many cases, vastly superior chemical durability.

Although true glasses are amorphous, there are a number of ways in which a glass can be microstructurally engineered through liquid–liquid phase separation or devitrification (crystallization). The *pièce de résistance* for glass is the plethora of forming modalities, which afford it flexibility on a par with metals. This chapter introduces some of the unique features of glass properties, forming structure, and composition that can be useful to the biotechnologist or health professional.

1.2 MAKING GLASS

Most commercial glasses are made by mixing raw materials of the oxides, carbonates, nitrates, or sulfates and heating to high temperature (for silicates, usually about 1450 °C) for a period of time, creating a melt.

The melt is then poured and cooled to create a solid glass. The reasons that the raw materials are not all oxides are numerous: the melting points of some of the alternatives are lower and they do in fact decompose to the oxides, releasing gases, which sweep other gases out of the melt and ultimately help to stir the melt to mix it. The latter is called fining and helps to ensure that the material is homogeneous. Most commercial glasses are melted in large gas furnaces in huge tanks made from refractory materials, except for some very special glasses that require ultra-high purities. For example, highly reactive (bioactive) glasses that will be made into medical devices have to be melted in platinum crucibles, which puts their manufacturing costs up significantly. Silica cannot be made by simply melting, since the temperatures required are too high, so silica will be addressed separately. The real difference in glass comes in the forming of the glasses.

How the glass is poured determines the shape of the glass obtained. A required shape can be made by pouring into a mould, often made of preheated graphite. Glass powders can be made by pouring the melt into water, creating a frit (see Figure 1 in colour section), which can then be ground. Fibers can be produced by drawing strands from the melt, which can be done at a controlled rate using a rotating drum.

Flat (window) glass for the construction and automotive industry is made using the float glass process, where the melt floats on molten tin that is kept in a reducing atmosphere to ensure no oxidation. The glass comes out almost perfectly flat under those circumstances. Not all glasses can be floated on molten tin owing to the viscosity–temperature relations, so some may be drawn up by sheet from the melt or they may be fed into an overflowing trough, which results in a highly pristine and thin flat glass; the process is known as a fusion draw.

Regardless of the forming processes used for the glasses, there is always a danger due to the nature of glass cooling, which dictates that different cooling rates will yield different structures. The bonding together as a solid of any two structures that do not have the same volume is going to result in stress; such stress can ultimately destroy the glass. The problem is that, if the thermal conductivity of the glass were good, then the cooling would spread more evenly. Unfortunately, this is not the case, and frozen-in stresses are commonplace. Hence the glasses are typically annealed; that is, they are either cooled or heated to a point where the said stresses have enough kinetic energy to relax, and then the glass is subsequently cooled slowly so as not to reintroduce further stresses. If the glasses are not annealed, then the competing stresses will likely lead to crack growth, which results in catastrophic failure. The mechanical properties of glass are probably its weakest link, although as we will see later there are ways of making it strong.

The theoretical strength of most commercial glasses is very high owing to the fact that they typically exhibit a mixture of strong covalent and ionic bonding. However, most glasses exhibit brittle behavior in which fracture occurs via the propagation of cracks that typically originate from a flawed surface. Therefore, glasses seldom exhibit anything close to their theoretical strength. The reason is that the sharp crack tips concentrate the stress applied to the glass, raising the stress locally way above the stress required to break bonds. Glasses without surface flaws are very strong. However, it takes very little to cause a surface flaw and weakening of the glass. A crack tip could result just from touching the surface. In addition, since the mechanical properties are so flaw-dependent, the "strength" of a glass product is statistically varying, and thus the engineering strength is determined by the weakest samples rather than by an average. Strengthening of glass involves the inhibition of crack growth either by blocking the path of the crack by an interface or by forcing the crack to be under compression rather than in tension. The latter can be achieved by using the property of glass mentioned earlier, namely that the structure of the glass is determined by its cooling rate. Tempered glass is purposely rapidly cooled on the outside to freeze-in a high-volume structure. As the glass on the inside cools slowly, it will try to relax to a lower-volume structure. However, the glass on the surface is already solid and is therefore pulled into compression as the inside glass tries to contract. The compressive surface layer serves to close the surface flaws, preventing them from propagating. It is not impossible to break tempered glass, but the compressive external layer has to be breached first, usually by a sharp object. Once through to the inner layer, which is in tension, immediate widespread catastrophic failure occurs. Tempered glass, because of the temperature gradients required, typically needs to be thick. Strong thin glass can be achieved by putting the surface into compression chemically. If solid commercial glass containing sodium ions is put in a bath of molten potassium salt, the sodium ions will come out into the salt and the potassium ions will diffuse into the glass. If this is done at a temperature above which the glass will not relax, then the large potassium ions forcing their way into the smaller sodium sites puts the surface into compression, similar to tempering. The rate at which this occurs and the ability to have a glass that does not relax or devitrify at the temperature of a molten salt bath depends upon the composition. Nevertheless, strengths have been improved almost a factor of 10 using these methods.

Removing surface flaws by dissolving a small amount of surface can increase strength greatly until new flaws are introduced. Mechanical properties are also dependent on morphology; for example, the bending strength (and flexibility) of a glass increases as fiber diameter decreases. This is why endoscopes can be used in medicine. Fiber optics rely on the optical properties of glasses and the fact that thin fibers can bend without breaking as long as they do not get surface flaws. Surface flaws are prevented by coating the glass fiber with a polymer sheath as soon as the fiber is pulled from the melt. The optical fibers can then be used for keyhole surgery.

1.3 HOMOGENEITY AND PHASE SEPARATION

We tend to think of glass as a very homogeneous material, but it can be inhomogeneous to a greater or lesser extent. In some cases, for commercial glasses widely used in the biotechnology industry, the inhomogeneity is on quite an appreciable scale and does affect the behavior of biological moieties quite significantly. Even silica, the most homogeneous of glasses, is inhomogeneous on a scale smaller than the wavelength of light. The fundamental random nature of glass means that there are small fluctuations in density. However, there is evidence to suggest that, even in commercial soda–lime–silica glass, the basis for most of our commercial compositions, there is some regionality of the structure consisting of compositional variations. We know for certain that most borosilicate glasses (such as those used to hold high-temperature or corrosive liquids, e.g. Pyrex®) exhibit compositional fluctuations on a large scale that can ultimately grow, when heat-treated, manifesting as a fogginess in the glass. This is liquid–liquid phase separation. The fogginess results from the scattering of light from the interfaces between the compositional fluctuations, which occurs because the refractive index undergoes a sudden change. These types of microstructures, if nurtured correctly, can ultimately help in making porous glasses. Phase separation is more the rule than the exception in glasses, but controlled phase separation usually has to be metastable, that is, developed in the solid glass. There are two types of immiscibility that occur: the formation of a droplet phase in a matrix; and spinodal decomposition, which results in an intertwined microstructure of two compositions. It is the latter that can result in very well-defined porous glasses (see Chapter 6). In addition to microstructural engineering by phase separation, where the phases are both glass, one can also develop a phase within the glass that is crystalline. These are called glass-ceramics and are used very widely in medicine and biotechnology (see Chapter 7). The chief advantages of devitrifying glasses (devitrifying means nucleating and growing a crystalline phase in an amorphous glass) are usually manifest in their mechanical properties, where crack propagation can be inhibited by the presence of the glass-crystal interface, and also in thermal expansion, where growth of a crystal phase can help control the net thermal expansion of a glass.

1.4 FORMING

Glass affords enormous flexibility in terms of forming, that is, making different shapes. This is in part due to the ability to cast glass in near net shapes. Commercial glasses tend to be categorized as flat, container, fiberglass, and specialty glass, where "specialty glass" is more or less everything else. Flat glass is generally made three ways: floating on a bath of molten tin (possible for a limited range of compositions, but very efficient and perfectly flat, and usually used for windows, etc.); pulling up from a melt through a rectangular hole; and allowing glass to overflow from a trough (e.g. used for liquid-crystal displays). Automotive glass, even though rarely "flat," comes under this category – it starts out as flat, but is slumped (heated slowly over a shaped mold) to get the aerodynamic shapes required for modern vehicles.

Container glasses are usually soda–lime–silica glasses, except for pharmaceutical glasses, which are type I (borosilicate, chemically resistant), type II (dealkalized soda–lime–silicate glass), or type III (low-alkali soda–lime–silicate glass). Most soda–lime–silicate containers are very durable in water and slightly acidic solutions but are attacked by basic solutions. Dependent upon the contents, the glass may be doped to be a certain color if the products within are sensitive to light. Containers are made in an automated process that blows them into a mold. Another forming method that has recently been finding new application in the field of biotechnology and medicine is the processing of glass microspheres. These microspheres are a fraction of the thickness of a human hair and can be injected into the bloodstream or their large controlled surface area can be used to stabilize biological molecules. Glass can also be made into fibers, both as a structural material and as an optical material. We are used to the former as being a component that keeps our homes well-insulated from heat and cold, and also as "fiberglass," which is a composite of a polymer matrix reinforced with glass fibers for a lightweight but strong material. The glass increases the stiffness of the polymer and the polymer matrix helps prevent the fibers from cracking.

However, there are applications for glass fibers in filtration technologies within the realm of biotechnology and even as a biocompatible fiberglass resin for repair in medicine. Optical fibers, such as those used in endoscopes, tend to be somewhat thicker than the structural fibers and have a more complex structure involving a high-refractive-index core that carries the light and a low-refractive-index cladding that confines the light to the core and protects the glass. Glass can be fiberized whether from the melt (structural fibers) or by heating and stretching a preform that has the required optical profile (optical fiber). Now micro- and nanoscale fibers can be generated by electrospinning and laser spinning techniques (Chapter 3). The fine-scale two-dimensional meshes and three-dimensional constructs mimic the fine scale of natural cell matrices and can be manipulated easily to fit into bone defects or used as membranes to guide tissue growth.

1.5 GLASSES THAT ARE NOT "MELTED"

High-purity glasses and glasses that would otherwise be difficult to form from the melt due to the high temperatures required can be made by other routes. One of the more popular ways is by vapor deposition. Silicon tetrachloride is a liquid at room temperature, which can be boiled and reacted with high-purity oxygen to form silica:

$$SiCl_4(g) + O_2(g) \rightleftharpoons SiO_2(s) + 2Cl_2(g)$$

The resulting highly pure silica deposits on a substrate as a low-density soot, which can be consolidated into vitreous silica by heat treatment. It is also possible to make doped silica glasses by this route, either by adding a component through reaction of a vapor phase with oxygen or by doping with a solution. Silica is a very important glass in biotechnology and medicine, as it is possibly the most durable glass and it is relatively straightforward to carry out surface engineering.

An alternative way of making high-purity and highly refractory glass is to use the sol-gel process (Chapter 3). Silica can easily be made but there are also some glasses that cannot be made by melting but can be made using a sol-gel method, for example, titania and simplified SiO_2–CaO compositions (sodium is not needed as there is no melting and therefore no need to lower the melting point). The sol-gel method is a chemical reaction method in which an organometallic compound is hydrolyzed (reacted with water). The resulting species then subsequently undergo a condensation reaction to form the bridging bonds of the glass network.

There will be more about this in Chapter 3. One is not restricted to simple glasses – multicomponent glasses can also be made. There are several other advantages to sol-gel processing aside from being able to obtain high-purity glasses, but two chief advantages are: (i) it is possible to engineer porosity (from nanoscale to microscale) in the glass; and (ii) since the glass does not have to be taken to a high temperature to solidify, one can incorporate temperature-sensitive molecules.

1.6 EXOTIC GLASS

In the glass world, the term "exotic glasses" usually refers to glasses that are not common silicates, but it can also mean glasses that have a novel application. In theory, one can make a glass from any element or compound if one cools it fast enough. There are many inorganic glasses that have found applications in biotechnology or medicine. Phosphate glasses are safely soluble in the body and are discussed in Chapter 4. Glasses such as borosilicates (Pyrex) are known for their chemical stability. However, silica-free glasses made from borate networks are soluble and have shown great potential in tissue regeneration, particularly in wound healing, such as healing diabetic ulcers. Deemed exotic even though they are a silicate are Bioglass® type compositions, which are explained in Chapters 2, 3, 5, 6, 8, 9, 11, and 12. There is a whole class of glasses made from fluoride compounds that can transmit infrared light of longer wavelength than can the silicates. Chalcogenide glasses made from combinations of elements from group 16 of the Periodic Table, such as arsenic, antimony, tellurium, selenium, and sulfur, also can transmit infrared light at longer wavelengths than the fluorides. The applications for these glasses in medicine have mostly been associated with their role as a conduit for infrared laser light. Not widely used yet, but likely to have a huge impact on medicine and biotechnology in the near future, are amorphous metals, which have interesting magnetic properties and outstanding mechanical properties and corrosion resistance.

1.7 SUMMARY

The take-home message for biotechnologists and practitioners of medicine from this introductory chapter should be that a glass is not a glass is not a glass. By this I mean that there is no one material "glass" and that care has to be taken to characterize what glass one has.

That said, glass has the flexibility and potential for a vast number of applications in biotechnology and medicine, many of which have not yet been thought of. For further reading, see the three books listed following this chapter. For reading on glass applications for biotechnology and medicine, please enjoy the rest of this book.

FURTHER READING

[1] Doremus, R. (1994) *Glass Science*, 2nd edn. Salt Lake City, UT: Academic Press.
[2] Shelby, J.E. (2005) *Introduction to Glass Science and Technology*, 2nd edn. Cambridge: Royal Society of Chemistry.
[3] Varshneya, A.K. (2006) *Fundamentals of Inorganic Glasses*, 2nd edn. Salt Lake City, UT: Academic Press.

2

Melt-Derived Bioactive Glass

Matthew D. O'Donnell

Department of Materials, Imperial College London, London, UK

2.1 BIOGLASS

2.1.1 Introduction to Bioglass

Bioglass®, also known as 45S5 bioactive glass, was the name given by
the University of Florida to the first material that was found to form a
bond with bone. It was developed in the late 1960s by Larry Hench after
he was challenged by a US Army Medical Officer to develop a material
to fill bone voids that would not be rejected by the body [1]. Bioglass is
a soda–lime–phosphosilicate (Na_2O–CaO–$P_2O_5 \cdot SiO_2$) glass that reacts
in the physiological environment. So the glass is a silica network with
other components, similar in many respects to window glass (Chapter 1)
except that there is a lot more calcium and a lot less silica in the glass.
Window glass also does not contain phosphate.

When implanted, the glass degrades slowly and dissolution products
stimulate progenitor cells to differentiate down a bone cell (osteoblast)
pathway by stimulating genes associated with osteoblast differentiation [2]. This phenomenon is called osteoinduction. The glass bonds to

Bio-Glasses: An Introduction, First Edition. Edited by Julian R. Jones and Alexis G. Clare.
© 2012 John Wiley & Sons, Ltd. Published 2012 by John Wiley & Sons, Ltd.

existing bone (osseintegrates) and encourages new bone growth along its surface (osteoconductive). Bone bonding is due to the formation of a calcium phosphate (hydroxycarbonate apatite, HCA) layer on the glass surface, which happens in the first few hours after implantation. The HCA layer has a similar composition to natural bone mineral. Collagen fibrils from host bone are thought to interact, via cellular processes, with the fine (nano)scale topography of the HCA layer, creating a strong bond.

The main application for Bioglass is as a synthetic bone graft for orthopaedic and periodontal use. In both these applications, the glass is used to regenerate and heal bone defects (holes in bone) that result usually from trauma or tumour/cyst removal. Periodontal defects are usually defects in the jaw bone around the tooth root.

Currently, orthopaedic surgeons use autografting, which is the transplanting of bone from another part of the body (a donor site) to the defect site. The donor site is usually the pelvis, or for spinal surgery it could be bone spurs from the vertebrae. In either case, the amount of bone is limited in quantity, and harvesting the bone in turn leaves a defect needing repair at the donor site. The donor site defect causes substantial pain to the patient and lengthens the recovery time. Complications arising at the donor site can lead to the need for further (revision) surgery.

Most recently, the original Bioglass can be found as the active ingredient in a leading toothpaste for reducing sensitivity in teeth. Sensodyne Repair & Protect (GlaxoSmithKline, UK) is available in more than 20 countries and is said to be 'powered by NovaMin' (see Figure 2 in colour section). NovaMin® (GlaxoSmithKline, UK) is a fine particulate of Bioglass that reduces pain felt in teeth when people eat or drink hot or cold items. Pain is felt when the dentine is exposed, perhaps as a result of receding gums. Dentine contains tubules that lead to nerve endings (Chapter 11). When hot or cold fluid flows down these tubules, pain is felt. Previous toothpastes for sensitive teeth have used potassium salts that temporarily desensitise the nerve endings. Clinical trials based on brushing with NovaMin indicate that the glass particles release ions and raise pH, causing precipitation of hydroxyapatite-like (HA) mineral over the tubule ends, blocking the tubules. The mineral has a similar composition to tooth enamel and therefore is beneficial to the tooth. Several patents have also been filed from a range of leading international researchers with the aim of improving further on the performance of bioactive toothpaste.

Table 2.1 Selected properties of Bioglass [3–6].

Property	Value
Density	$2.7\,g/cm^3$
Glass transition temperature	$538\,°C$
Crystallisation onset temperature	$677\,°C$
Melting temperature	1224 and $1264\,°C$
Thermal expansion coefficient	$15.1 \times 10^{-6}\,°C^{-1}$
Refractive index	1.59
Tensile strength	$42\,MPa$
Young's modulus (stiffness)	$35\,MPa$
Shear modulus	$30.7\,GPa$
Fracture toughness	$0.6\,MPa\,m^{1/2}$
Vickers hardness	$5.75\,GPa$

2.1.2 The Materials Properties of Bioglass

Table 2.1 shows various thermal, physical, mechanical and chemical properties of Bioglass. Bioglass and other bioactive glasses have poor mechanical properties, with a lower tensile strength and higher modulus compared to cortical bone ($50–150\,MPa$ and $7–30\,GPa$, respectively), which means that they cannot be implanted into load-bearing bone defects alone. So, they could not be used alone to regenerate a full-thickness (segmental) bone defect – metallic fixation is required in those applications to take the cyclic load. Therefore, bioactive glasses and other bioactive ceramics are more often used to repair defects that are surrounded by host bone.

However, bioactive glasses can be incorporated in composite structures and the glass can be formed into scaffolds or fibres to improve mechanical properties and speed the formation of biomimetic apatite after implantation.

2.1.3 Mechanism of Bioactivity and Effect of Glass Composition

The original Bioglass is a quaternary soda–lime–phosphosilicate glass. The composition is shown in Table 2.2. The silica content is relatively low compared to window and container glass, to aid dissolution. The calcium and sodium content is high. This lowers the melting temperature but importantly also aids dissolution. On degradation, owing to the high content of alkali-metal and alkaline-earth ions, the pH of the

Table 2.2   Bioglass (45S5) composition in molecular and weight percentages.

Composition	SiO_2	Na_2O	CaO	P_2O_5
Mol%	46.13	24.35	26.91	2.60
Wt%	45.00	24.50	24.50	6.00

solution rises. Increases in pH can cause the calcium and phosphate ions that are naturally present in body fluid to come out of solution. Phosphorus and calcium are also released into the physiological environment by the glass. Accompanied with the pH rise, this drives the formation of HCA at the glass surface, which is chemically similar to the inorganic part of bone.

The mechanism for dissolution and bone bonding of a bioactive glass proposed by Larry Hench is a multi-stage process involving the following steps [7, 8]:

1. rapid ion exchange of alkali-metal cations (e.g. Na^+, Ca^{2+}) with H^+ from body fluid;
2. loss of soluble silica (effectively silicic acid), leaving behind –Si–OH bonds;
3. condensation and repolymerisation of the Si–OH bonds to create a silica-rich (cation-depleted) layer;
4. migration of Ca^{2+} and $[PO_4]^{3-}$ groups from inside the glass and from the body fluid, forming an amorphous calcium phosphate layer that grows on the silica-rich layer surface; and
5. crystallisation of the amorphous layer by incorporation of OH^- and $[CO_3]^{2-}$ from solution to form HCA.

After the formation of crystalline HCA, the following steps were proposed by Hench [8] to occur:

6. biological moieties absorbed in HCA layer;
7. macrophage action;
8. attachment of stem cells;
9. differentiation of stem cells;
10. generation of matrix; and
11. crystallisation of matrix.

A compositional diagram of the SiO_2–Na_2O–CaO system is shown in Figure 2.1 [8]. All compositions have a fixed 6 wt% of phosphate

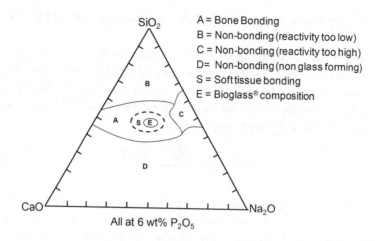

Figure 2.1 Bioactivity map of compositions in the SiO_2–Na_2O–CaO system (6 wt% of P_2O_5) showing regions of bioactive response.

and all those in region A bond to bone. Glasses such as window and container glass would fall in region B and are effectively bio-inert and implantation would result in scar tissue encapsulation. This is a result of the silica network being too dense and resistant to reaction with body fluid. Glasses in region C are the opposite and degrade too rapidly (<30 days), as the silica content is too low and the soda content too high, leading to a highly disrupted network. Region D contain compositions with such low silica that either a glass network cannot form or crystals form. Glasses within the dashed line (region E within region A) strongly adhere to collagenous (soft) tissue as well as the inorganic part of bone. 45S5 Bioglass falls within regions A and E.

The reason behind the composition affecting dissolution rate in body fluid is the connectivity of the silicate network. A highly connected network containing a large proportion of bridging oxygen bonds (silica tetrahedra covalently bonded to other silica tetrahedra via –O–Si–O– bonds) is durable in body fluid. High connectivity is created by high silica content, such that melt-derived glasses with more than 60 mol% silica are not bioactive. Connectivity is lowered by adding the network-modifying (network-disrupting) cations such as sodium and calcium. The network connectivity (NC) (number of bridging oxygen bonds per silicon atom) of 45S5 Bioglass is approximately 2, making the glass susceptible to degradation. Chapter 5 discusses the atomic structure of bioactive glasses in more detail.

2.2 NETWORK CONNECTIVITY AND BIOACTIVITY

The NC of an inorganic glass is the number of bridging oxygens per $[SiO_4]^{4-}$ tetrahedron. As this is a measure of the degree of cross-linking and connectivity of the silica, it is a useful tool for predicting glass properties from the composition, including bioactivity [9]. NC can be calculated by the following equation:

$$NC = 2 + \frac{BO - NBO}{G} \qquad (2.1)$$

where BO is the total number of bridging oxygens per network-forming ion, NBO is the total number of non-bridging oxygens per network modifier ion and G is the total number of glass-forming units. For an SiO_2–Na_2O–CaO–P_2O_5 system, where P is assumed to enter the glass network (i.e. formation of P–O–Si bonds), NC can be calculated as follows:

$$NC = 2 + \frac{[(2 \times SiO_2) + (2 \times P_2O_5)] - [(2 \times Na_2O) + (2 \times CaO)]}{[SiO_2 + (2 \times P_2O_5)]}$$

$$(2.2)$$

For 45S5 Bioglass ($46.13SiO_2$–$24.35Na_2O$–$26.91CaO$–$2.6P_2O_5$ mol%), NC is equal to:

$$NC = 2 + \frac{[(2 \times 46.13) + (2 \times 2.6)] - [(2 \times 24.35) + (2 \times 26.91)]}{[46.13 + (2 \times 2.6)]}$$

$$= 1.90 \qquad (2.3)$$

In melt-derived bioactive glass systems, phosphate typically forms $Q^0[PO_4]^{3-}$ units rather than forming part of the silicate network. This phosphate complex requires three positive charges from other cations to charge-balance itself [10]. The lower the NC, the more susceptible a glass is to degradation, and therefore the more bioactive it is likely to be. Therefore, the calculation must be modified, and NC can be calculated as follows:

$$NC' = 2 + \frac{[2 \times SiO_2] - [(2 \times Na_2O) + (2 \times CaO)] + (2 \times P_2O_5 \times 3)}{[SiO_2]}$$

$$(2.4)$$

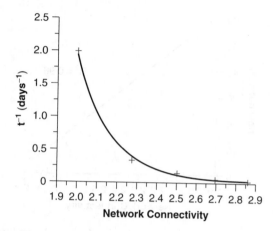

Figure 2.2 Effect of NC on rate of apatite formation in simulated body fluid on phosphate-free soda. (Plotted from data in Ref. [11].)

For 45S5 Bioglass, NC′ is equal to:

$$NC' = 2 + \frac{[2 \times 46.13] - [(2 \times 24.35) + (2 \times 26.91)] + (2 \times 2.6 \times 3)}{[46.13]}$$

$$= 2.22 \tag{2.5}$$

Glasses that have NC (rather than NC′) greater than 2.4 are not likely to be bioactive. Figure 2.2 shows how NC affects HCA formation on SiO_2–CaO–Na_2O glass discs in simulated body fluid (SBF). Above NC = 2, the rate of HCA layer deposition decreases rapidly.

This trend can also be seen for *in vivo* bone ingrowth for particles implanted in a bone defect (Figure 2.3). The percentage of new bone formed (osteoconduction) drops off rapidly for glasses where NC > 2.

Other properties can be predicted from NC, such as tendency for crystallisation, as glasses with NC > 2 will generally have a higher energy barrier to overcome for crystallisation than glasses with NC < 2 owing to the stabilising effect of cross-linking the silicate chains.

2.3 ALTERNATIVE BIOACTIVE GLASS COMPOSITIONS

Research groups worldwide have developed a large number of bioactive glass compositions using the 45S5 composition as a guide or starting

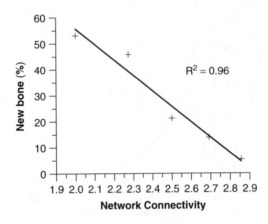

Figure 2.3 Effect of NC on osteoconduction *in vivo*. (Plotted from data in Ref. [11].)

point. Because glass is a random arrangement of units, such as the silica tetrahedra, there is space for a large number of therapeutic ions. The ions can be incorporated into the chemistry without significant alteration of properties. 45S5 crystallises rapidly when it is heated above T_g. This presents a problem if engineers want to process the glass in its more fluid form. This could be sintering (fusing) particles together to make large pieces of glass, porous constructs or scaffolds (Chapter 12) or it could be coating metal implants with glass (Chapter 8). Another problem with making coatings using Bioglass is that the thermal expansion coefficients (TECs) of the glass and metal (e.g. titanium or cobalt–chrome alloy) do not match, which causes the glass coating to shrink away from the metal surface during the coating process. Ions such as zinc, magnesium and boron can be added to the glass to stabilise the working range to enable viscous flow sintering and to alter the TEC to match those of metal alloys for coatings [5, 12]. Potassium and fluorine can be added for dental application when slow release of fluoride ions is desired. Silver can also be added for bactericidal action. Strontium has been shown to be beneficial for healthy bone growth, and glass particles containing strontium are now commercially available as StronBone™ (RepRegen Ltd, UK). The release of these ions can be controlled by altering the NC and particle size of the glasses. Table 2.3 summarises some important bioactive glass compositions reported in the literature.

Figure 2.4 shows the effect of the addition of strontium on the thermal properties using differential thermal analysis (DTA) in a bioactive glass series [4]. DTA measures thermal changes as the temperature is increased,

Table 2.3 Non-45S5 bioactive glass compositions in weight percentages.

Name	ID	SiO_2	Na_2O	CaO	P_2O_5	SrO	K_2O	MgO
NovaBone	45S5 Bioglass	45.0	24.5	24.7	6.0	–	–	–
BonAlive	S53P5	53.0	23.0	20.0	4.0	–	–	–
	13–93	53.0	6.0	20.0	4.0	–	12.0	5.0
	Sr10	44.1	24.0	21.6	5.9	4.4	–	–
	ICIE1-10Sr	48.2	26.5	18.9	2.5	3.9	–	–

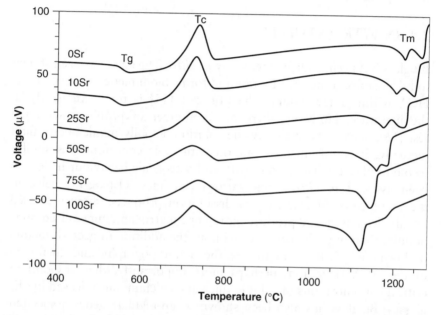

Figure 2.4 Effect of substituting calcium with strontium on thermal properties using DTA on a series of bioactive glasses. (Reprinted with permission from [4] Copyright (2010) Royal Society of Chemistry.)

so an upward peak is an exothermic transition (bond-forming) and a trough is an endothermic transition (bond-breaking). The characteristic glass transition temperature (T_g) and the melting temperature (T_m) both decrease with increasing strontium [13]. The crystallisation onset temperature (T_c) is an exception, in that it approaches a minimum at 50% strontium substitution (perhaps approaching a metasilicate eutectic point), then increases again in the Sr-rich glasses.

Magnesium is particularly effective at widening the sintering window, but also affects bioactivity. Magnesium is typically considered a network modifier in silicate glasses, but it can partially act as an intermediate oxide in highly disrupted bioactive glasses owing to its charge-to-size ratio (field strength). When magnesium is substituted for calcium in bioactive glasses, approximately 14% of the magnesium forms $[MgO_4]^{2-}$ units, which, like the Q^0 phosphate units, remove network modifiers (e.g. Ca^{2+}) from the silicate network, increasing the connectivity of the silicate network and reducing its solubility and reactivity [14]. Therefore, above a critical concentration, magnesium should be considered a network former in NC calculations, but that threshold will depend on the rest of the glass composition.

2.4 *IN VITRO* STUDIES

Bioglass has been studied extensively *in vitro*. The glass has been shown to be bioactive using the SBF test and forms biomimetic nanocrystalline HCA within a few hours. The SBF test [15] is used extensively in the literature. However, there are a number of problems with the test [16], such as how closely the solution actually mimics body fluid (actually, it is an aqueous solution with ionic concentrations similar to blood plasma). The cytotoxicity and osteogenic behaviour have also been assessed *in vitro* using various cell lines. Osteogenic cells, in the presence of Bioglass, or its dissolution products, show increased metabolic activity and produce more bone matrix and mineralised bone nodules than other bioceramics without the addition of growth factors or hormones. Cells seem to like the nanotopography and chemistry of the HCA layer and, perhaps more importantly, they respond to critical concentrations of soluble silica and calcium ions released by the glasses. Bioglass has also been shown to up-regulate genes associated with bone formation, including *cMyc*-responsive growth-related gene, cell cycle regulators, apoptosis regulators, cell surface receptors and extracellular matrix regulators [17]. The effect is dose-dependent. This is important, as too many ions would raise the pH too high and start damaging cells. It is a challenge to engineers, surgeons and biologists to work together to understand exactly how different cells respond to the different topographies and chemistries of materials.

2.5 *IN VIVO* STUDIES AND COMMERCIAL PRODUCTS

2.5.1 Animal Studies

Animal studies are needed before new medical devices can be implanted in people. To minimise the number of animals used and to obtain

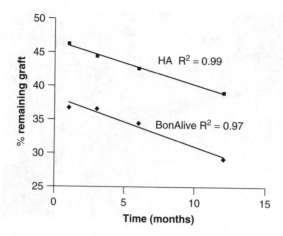

Figure 2.5 Resorption of BonAlive bioactive glass (BG) and hydroxyapatite (HA) *in vivo*. (Plotted from data in Ref. [19].)

reliable results, effective models are needed. The definitive animal model for investigating the efficacy of bioceramics in bone regeneration was developed by Oonishi *et al.* [18] using the rabbit femoral condyle (the leg bone just above the knee). They compared the resorption of Bioglass to that of synthetic HA and an apatite wollastonite glass-ceramic (AWGC) *in vivo*. After 12 weeks, 30–50% of the bioactive glass particles had resorbed away and were shown to stimulate more bone growth than AWGC and HA. Figure 2.5 shows the resorption of a bioactive glass (BonAlive) and HA over a one-year period after implantation at two different sites in rabbit frontal sinuses [19]. Approximately 30% of the glass implants remained after 12 months (the difference replaced by new bone). This is a good time frame for surgeons dealing with bone repair, as it is important that the resorption of the implant does not happen before the bone has the chance to regenerate.

In one of the only studies to compare the *in vitro* and *in vivo* behaviour of bioactive glasses, Fujibayashi *et al.* [11] studied the bioactivity of a series of ternary SiO_2–Na_2O–CaO glasses (i.e. phosphate-free). In this case, the *in vitro* bioactivity (rate of HCA formation in SBF) decreased as the silica content increased, which correlated well with *in vivo* response.

To investigate excretion of the dissolved silica from the body, Bioglass excretion rates of silica were studied following implantation of Bioglass in rabbit muscle [20]. The average excretion rate of silicon in urine was 2.4 mg/day (well below saturation). Excretion rates of silicon fell

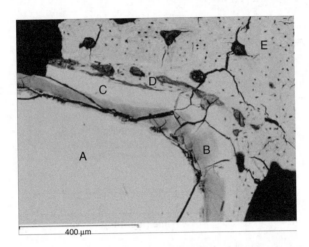

Figure 2.6 Backscattered electron micrograph of implanted glass ICIE1Sr10 in osteoporotic rat femur: A = glass, B = Si gel layer, C = HCA layer, D = existing bone and E = new bone growth into marrow cavity. (Adapted with permission from [21]. Copyright (2009) Imperial College London.)

to the basal level after 19 weeks. There was no elevated concentration of silicon in the organs (brain, heart, kidney, liver, lung, muscle and thymus) at 24 weeks. Figure 2.6 shows an electron microscope image of a strontium-containing bioactive glass cylinder that was implanted in an osteoporotic rat femur after four weeks. The image nicely shows the different regions of a bioactive glass that has reacted with blood and bonded to bone: a silica gel layer formed at the surface of the glass (due to ion exchange of Na^+ and Ca^{2+} ions from the glass with H^+ from the blood), with an apatite layer in between this and the existing bone. This shows evidence of osteointegration with direct bone contact. New bone has formed and is growing into the defect and marrow cavity area, indicating osteoconduction.

2.5.2 Human Clinical Studies and Commercial Products

As Larry Hench writes in the Foreword, the first Bioglass product was a middle ear prosthesis [22]. Although it worked, it was only available in two sizes of cone-shaped implants and every clinical case was slightly different, requiring a device that could be shaped by the surgeon. This limited the commercial success of the device and is also a problem for all bioceramic implants: surgeons want to be able to shape and/or inject the device.

Figure 2.7 Scanning electron microscope image of Bioglass particles with similar morphology and size range as PerioGlas and NovaBone products. Scale bar is 200 μm.

The vast majority of human clinical trials on Bioglass have been performed using PerioGlas®, which is a particulate for periodontal tissue repair. Figure 2.7 shows an electron microscope image of Bioglass particles that are used in PerioGlas. Surgeons usually mix the particles with blood from the defect so that the powder becomes a putty-like material that can be pushed into the defect. The blood also acts to introduce cells and natural growth factors into the defect to further promote regeneration. In comparative studies, PerioGlas performed as good as, or better than, the following controls: open debridement [23], allograft [24], membrane and HA [25]. A number of other dental and maxillofacial clinical studies have shown positive results, with the bioactive glass promoting remineralisation at the defect site, including repair of orbital floors beneath the eye and sinus floors.

NovaBone® (NovaBone Products LLC, USA) is a particulate very similar to PerioGlas that is approved for orthopaedic bone defect repair (see Figure 2 in colour section). It is also available in a putty form, where the particles are dispersed in a putty-like matrix. A slight modification of the Bioglass composition called BonAlive® (Vivoxid, Finland) has been developed in Finland as a bone graft substitute [19]. StronBone is another particulate similar to NovaBone, except that it contains some strontium, which is slowly released from the glass after implantation. Bioglass has been incorporated in a composite of collagen, Bioglass and tricalcium phosphate bone void filler called Vitoss-BA® (Orthovita, USA), which shows higher new bone strength at 12–52 weeks after implantation compared to the Bioglass-free material. Orthofix (Texas, USA) have also developed a composite bone void filler called Origen® made from Bioglass, demineralised bone matrix (DBM) and gelatin.

Fine particles of Bioglass much smaller than those used in a bone graft are now incorporated in toothpaste as the product NovaMin, as described earlier [26]. The fine particulate cannot be detected by the user but also increases the surface area and hence solubility, which is necessary as the glass particles are not present in the saliva for long and they need to dissolve and raise pH to mineralise dentine tubules and reduce sensitivity. The toothpaste has low water content and is based on poly(ethylene glycol) (PEG) and glycerol to minimise the reactivity of the glass in the tube.

Patents have been filed in other areas such as using bioactive glasses in a range of personal care products (Schott) and in hair straightening formulations (L'Oreal).

REFERENCES

[1] Hench, L.L. (2006) The story of Bioglass. *Journal of Materials Science: Materials in Medicine*, 17, 967–978.

[2] Hench, L.L. and Polak, J.M. (2002) Third-generation biomedical materials. *Science*, 295, 1014–1017.

[3] Jedlicka, A.B. and Clare, A.G. (2001) Chemical durability of commercial silicate glasses. I. Material characterization. *Journal of Non-Crystalline Solids*, 281, 6–24.

[4] O'Donnell, M.D., Candarlioglu, P.L., Miller, C.A. *et al.* (2010) Materials characterisation and cytotoxic assessment of strontium-substituted bioactive glasses for bone regeneration. *Journal of Materials Chemistry*, 20, 8934–8941.

[5] Lopez-Esteban, S., Saiz, E., Fujino, S. *et al.* (2003) Bioactive glass coatings for orthopedic metallic implants. *Journal of the European Ceramic Society*, 23, 2921–2930.

[6] Lin, C.C., Huang, L.C. and Shen, P.Y. (2005) $Na_2CaSi_2O_6-P_2O_5$ based bioactive glasses. Part 1: elasticity and structure. *Journal of Non-Crystalline Solids*, 351, 3195–3203.

[7] Clark, A.E., Pantano, C.G. and Hench, L.L. (1976) Auger spectroscopic analysis of Bioglass corrosion films. *Journal of the American Ceramic Society*, 59, 37–39.

[8] Hench, L.L. (1991) Bioceramics – from concept to clinic. *Journal of the American Ceramic Society*, 74, 1487–1510.

[9] Hill, R. (1996) An alternative view of the degradation of Bioglass. *Journal of Materials Science Letters*, 15, 1122–1125.

[10] O'Donnell, M.D., Watts, S.J., Law, R.V. and Hill, R.G. (2008) Effect of P_2O_5 content in two series of soda lime phosphosilicate glasses on structure and properties – Part I: NMR. *Journal of Non-Crystalline Solids*, 354, 3554–3560.

[11] Fujibayashi, S., Neo, M., Kim, H.M. *et al.* (2003) A comparative study between in vivo bone ingrowth and in vitro apatite formation on Na_2O–CaO–SiO_2 glasses. *Biomaterials*, 24, 1349–1356.

[12] Brink, M., Turunen, T., Happonen, R.P. and Yli-Urpo, A. (1997) Compositional dependence of bioactivity of glasses in the system Na_2O–K_2O–MgO–CaO–B_2O_3–P_2O_5–SiO_2. *Journal of Biomedical Materials Research*, 37, 114–121.

[13] O'Donnell, M.D. and Hill, R.G. (2010) Influence of strontium and the importance of glass chemistry and structure when designing bioactive glasses for bone regeneration. *Acta Biomaterialia*, 6, 2382–2385.

[14] Watts, S.J., Hill, R.G., O'Donnell, M.D. and Law, R.V. (2010) Influence of magnesia on the structure and properties of bioactive glasses. *Journal of Non-Crystalline Solids*, 356, 517–524.

[15] Kokubo, T. and Takadama, H. (2006) How useful is SBF in predicting in vivo bone bioactivity? *Biomaterials*, 27, 2907–2915.

[16] Bohner, M. and Lemaitre, J. (2009) Can bioactivity be tested in vitro with SBF solution? *Biomaterials*, 30, 2175–2179.

[17] Xynos, I.D., Edgar, A.J., Buttery, L.D.K. *et al.* (2001) Gene-expression profiling of human osteoblasts following treatment with the ionic products of Bioglass 45S5 dissolution. *Journal of Biomedical Materials Research*, 55, 151–157.

[18] Oonishi, H., Hench, L.L., Wilson, J. *et al.* (2000) Quantitative comparison of bone growth behavior in granules of Bioglass, A-W glass-ceramic, and hydroxyapatite. *Journal of Biomedical Materials Research*, 51, 37–46.

[19] Peltola, M.J., Aitasalo, K.M.J., Suonpaa, J.T.K. *et al.* (2001) In vivo model for frontal sinus and calvarial bone defect obliteration with bioactive glass S53P4 and hydroxyapatite. *Journal of Biomedical Materials Research*, 58, 261–269.

[20] Lai, W., Garino, J., Flaitz, C. and Ducheyne, P. (2005) Excretion of resorption products from bioactive glass implanted in rabbit muscle. *Journal of Biomedical Materials Research Part A*, 75A, 398–407.

[21] Fredholm, Y.C. (2009) Development and characterisation of strontium-containing bioactive glasses and aluminium-free glass polyalkenoate cements. PhD thesis. Imperial College London, London.

[22] Merwin, G.E. (1986) Bioglass middle-ear prosthesis – preliminary report. *Annals of Otology, Rhinology and Laryngology*, 95, 78–82.

[23] Froum, S.J., Weinberg, M.A. and Tarnow, D. (1998) Comparison of bioactive glass synthetic bone graft particles and open debridement in the treatment of human periodontal defects. A clinical study. *Journal of Periodontology*, 69, 698–709.

[24] Lovelace, T.B., Mellonig, J.T., Meffert, R.M. *et al.* (1998) Clinical evaluation of bioactive glass in the treatment of periodontal osseous defects in humans. *Journal of Periodontology*, 69, 1027–1035.

[25] Oonishi, H., Kushitani, S., Yasukawa, E. *et al.* (1997) Particulate Bioglass compared with hydroxyapatite as a bone graft substitute. *Clinical Orthopaedics and Related Research*, 334, 316–325.

[26] Wefel, J.S. (2009) NovaMin: likely clinical success. *Advances in Dental Research*, 21, 40–43.

3

Sol-Gel Derived Glasses for Medicine

Julian R. Jones

Department of Materials, Imperial College London, London, UK

3.1 INTRODUCTION

Glass can be made using two processing methods: the traditional melt-quench route and the sol-gel route. In the melt-quench method, oxides are melted together at high temperatures (above 1300 °C) in a crucible and quenched in a mould or in water (Chapters 1 and 2). The sol-gel route is a low-temperature processing route where a solution containing the compositional components (sol) undergoes polymer-type reactions at room temperature to form a gel. The gel is a wet inorganic network, similar to a highly cross-linked short-chain inorganic polymer, which is then dried and heated, for example to 600 °C, to become a glass. The low-temperature process provides opportunities to make porous scaffolds (Chapter 12), incorporate drugs and growth factors (Chapter 12) and allow incorporation of polymers to make hybrid materials (Chapter 10).

The sol-gel process produces glasses with very different properties from conventional glasses, and their potential future applications are

Bio-Glasses: An Introduction, First Edition. Edited by Julian R. Jones and Alexis G. Clare.
© 2012 John Wiley & Sons, Ltd. Published 2012 by John Wiley & Sons, Ltd.

Figure 3.1 The versatility of the sol-gel process: (a) scanning electron microscope of a bioactive sol-gel glass disc made under acid catalysis; and (b) transmission electron microscope image of silica nanoparticles.

very exciting. For example, silicates and bioactive glasses can be made that are either mesoporous or nanoparticles (Figure 3.1). The difference in the sol-gel processing to produce the two materials is simply the catalyst used in making the sol.

This chapter will describe the sol-gel process, discuss its benefits and disadvantages over the melt-quench route, and describe some applications for the glasses. The most common sol-gel derived glasses for biomedical applications are silica-based (SiO_2) glasses, and therefore the emphasis in this chapter is on silica sol-gel processing. However, phosphate glasses have also been produced using the sol-gel process (Chapter 4). More details can be found in [1–3].

3.2 WHY USE THE SOL-GEL PROCESS?

The motivation for sol-gel processing is primarily the potentially higher purity and homogeneity and the lower processing temperatures associated with sol-gels compared with traditional glass melting or ceramic powder methods. The goal of sol-gel processing is to control the glass formation during the earliest stages of processing, producing uniquely homogeneous structures.

The purity and homogeneity of dense gel silica made by the hydrolysis and condensation of alkoxide precursors are superior to other silica glass processing methods. The ability to produce optics with nearly theoretical limits of optical transmission, lower thermal expansion coefficients and greater homogeneity, along with net shape casting, represent major advances resulting from sol-gel processing of monoliths. The process also allows the synthesis of glasses with a variety of pore structures from the nanometre to the millimetre range, which gives sol-gel glasses great potential in biomedical applications.

Fully densified sol-gel silica has physical properties and structural characteristics similar to high-grade fused silica but offers the advantages of near net shape casting, including internal cavities, and a lower thermal expansion coefficient of 0.2×10^6 cm/cm compared with 0.55×10^6 for conventional processing methods. Optically transparent porous gel silica optics have a density as low as 60% of fused silica and can be impregnated with up to 30–40 vol% of a second-phase optically active organic or inorganic compound.

Centrifugal deposition of 12–40 nm colloidal silica powders can be used to produce synthetic silica tubes used in the manufacture of optical telecommunication fibres. One of the main advantages of the sol-gel technique is very high accuracies for the diameter, cross-sectional area and wall thickness of the tubes.

3.3 SOL-GEL PROCESS PRINCIPLES

A common sol-gel process is the use of organic precursors that can undergo polymer-type reactions to form an inorganic network rather like a highly cross-linked amorphous inorganic polymer. Although modern compared to the discovery of traditional glass, which pre-dates history, the sol-gel process is over 150 years old. Work began in 1846 [4], with the discovery that the alkoxide tetraethyl orthosilicate (TEOS) $Si(OC_2H_5)$ reacted with water (hydrolysis) under acidic conditions. Polymer condensation reactions then occurred, forming an inorganic silica network (gel) with aqueous by-products. Drying the gel produced a glassy SiO_2. Unfortunately, there was little wider interest, as the drying times were in excess of a year to avoid the gels breaking into tiny pieces. However, 100 years later, the interest was rekindled. The processing of powders, fibres and coatings that did not crack during drying and the hydrolysis of TEOS became the basis for many sol-gel glass systems.

Gels that are dried under ambient conditions, or under carefully controlled heating, are referred to either as xerogels or simply as gels. However, it is still difficult to obtain crack-free monoliths under these conditions. When the pore liquid is removed as a gas phase from the interconnected solid gel network under hypercritical conditions (critical-point drying), the network does not collapse and a low-density aerogel is produced. Aerogels can have pore volumes as large as 98% and densities as low as 80 kg/m^3 [5]. Photographs of an aerogel are given in the colour section (see Figure 3 in colour section). Owing to their high porosity, aerogels are the lowest density solid materials in the world, making them very good for thermal insulation when sandwiched between glass plates and evacuated. NASA uses them to catch tiny meteorites by attaching them to the outside of their spacecraft. However, for biomedical applications, conventional sol-gels (xerogels) are of interest and they will be the focus of this chapter.

3.4 STEPS IN A TYPICAL SOL-GEL PROCESS

Each step of the acid-catalysed process will now be introduced. More detail can be found in [1]. A schematic of the process can be seen in Figure 3.2.

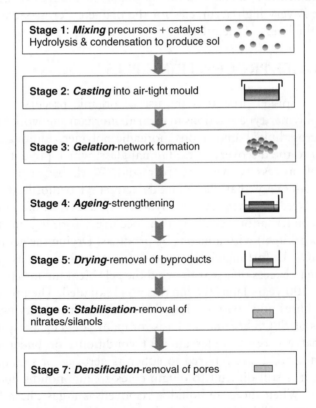

Stage 1: *Mixing* precursors + catalyst Hydrolysis & condensation to produce sol

Stage 2: *Casting* into air-tight mould

Stage 3: *Gelation*-network formation

Stage 4: *Ageing*-strengthening

Stage 5: *Drying*-removal of byproducts

Stage 6: *Stabilisation*-removal of nitrates/silanols

Stage 7: *Densification*-removal of pores

Figure 3.2 A flow chart of the seven stages for the sol-gel synthesis of a glass (acid catalysis).

The precursors for silicate-based glass production are usually alkoxides ($Si(OR)_4$, where R is CH_3, C_2H_5 or C_3H_7) and nitrates. Alkoxides are usually used as precursors of the network formers, such as silica and phosphate, whereas salts (e.g. nitrates) are usually used to introduce network modifiers such as calcium.

3.4.1 Stage 1: Mixing

Sols are created by hydrolysis of the precursors. Sols are dispersions of colloidal particles (diameters of 1–100 nm) in a liquid. The combination and quantities of the precursors mixed with water to create the sol will determine the composition of the glass. In the case of TEOS ($Si(OC_2H_5)_4$), it is insoluble in water, but a catalyst can trigger hydrolysis, where it reacts with water to effectively produce silica tetrahedra in water (Equation 3.1 and Figure 3.3).

$$
\begin{array}{c}
OC_2H_5 \\
| \\
C_2H_5O-Si-OC_2H_5 \\
| \\
OC_2H_5
\end{array}
+ 4(H_2O) \longrightarrow
\begin{array}{c}
OH \\
| \\
HO-Si-OH \\
| \\
OH
\end{array}
+ 4(C_2H_5OH)
\tag{3.1}
$$

The catalyst can be either an acid (e.g. HNO_3) or a base, and this will affect the final properties of the glass. The tetrahedra have surfaces of Si–OH groups and are floating in the water.

At pH < 2 (usually using a nitric acid catalyst) the tetrahedra will come together and the Si–OH groups will condense, forming O–Si–O bridging bonds, leaving water as a by-product (Equation 3.2).

$$
\begin{array}{c}
OH \\
| \\
HO-Si-OH \\
| \\
OH
\end{array}
+
\begin{array}{c}
OH \\
| \\
HO-Si-OH \\
| \\
OH
\end{array}
\longrightarrow
\begin{array}{c}
OH \quad OH \\
| \quad\; | \\
HO-Si-Si-OH \\
| \quad\; | \\
OH \quad OH
\end{array}
+ H_2O
\tag{3.2}
$$

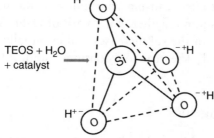

Figure 3.3 A silica tetrahedron, immediately after hydrolysis of a TEOS molecule.

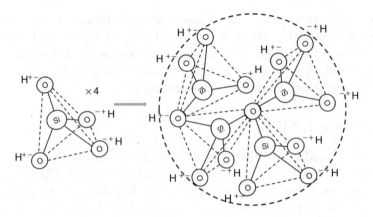

Figure 3.4 A schematic of silica tetrahedra forming a nanoparticle in a sol. The nanoparticle will coalesce with other similar particles and bonds will form between them to form a silica network. When it becomes rigid, it will be a gel.

The hydrolysis and polycondensation reactions occur simultaneously within the solution during mixing. When sufficient interconnected Si–O–Si bonds are formed in a region, they behave as colloidal (sub-micrometre) particles and a sol is formed (Figure 3.4). The first small particles that form are termed 'primary particles'.

3.4.2 Stage 2: Casting

The sol can be cast prior to gelation in air-tight moulds. The moulds must be made of material that will not adhere to the gel (e.g. polytetrafluoroethylene, PTFE). The gelation process can take several days at room temperature but can be tailored to the needs of the process.

3.4.3 Stage 3: Gelation

As mixing continues, agglomeration of the primary particles occurs and condensation continues, causing coarsening of the particles. As more and more particles join together, a network of O–Si–O bonds begins to form, the viscosity of the sol increases and eventually a gel (rigid silica network) is formed. The reaction continues until all the precursor is used up.

3.4.4 Stage 4: Ageing

After gelation, the gel is aged, either at room temperature or at slightly elevated temperatures, for example 60 °C, for several hours or days.

Importantly, the gel is sitting in its aqueous by-products during ageing. This process strengthens the gel by consolidation as the condensation reaction continues. Without the ageing stage, the gel will crack during drying.

3.4.5 Stage 5: Drying

The aim of the drying stage is to remove the by-products of the condensation reaction. As the water and alcohol evaporate, they leave behind an interconnected pore network. The pores have diameters in the nanometre range, typically 1–30 nm in diameter. Figure 3.4 shows that –OH groups are left behind on the surface of the particles.

3.4.6 Stage 6: Stabilisation

Although the drying process removes water, hydroxyl (OH) groups are left on the pore walls. In a silica sol-gel glass, the H^+ acts as a network modifier, disrupting the silica network and reducing network connectivity (Chapter 2 explains the concept of network connectivity and its relation to bioactivity). A chemically stable porous solid can be obtained by removing surface silanol (Si–OH) bonds from the pore network, usually by thermal processing. Thermal processing drives off the –OH groups, causing further formation of O–Si–O bonds. If other precursors, such as nitrates, have been used in the process, for example calcium nitrate to introduce calcium into the composition, the nitrates will also be removed during stabilisation. The usual method is to heat the dried gel to temperatures above 700 °C to produce a porous glass.

3.4.7 Stage 7: Densification

Some applications require a glass without the interconnected nanoporous network. This can be achieved by heating the glass at higher temperatures (~1000 °C for silica), causing densification (sintering) to occur. The pores are eliminated, and the density ultimately becomes equivalent to that of the melt-quenched glass. The densification temperature depends considerably on the morphology of the pore network, the surface area and the glass composition. Temperatures must be raised above the glass transition for sintering to occur, but they must be kept below the crystallisation temperature, otherwise a glass-ceramic will form.

3.5 EVOLUTION OF NANOPOROSITY

The main difference that sets sol-gel derived glasses apart from melt-quenched glasses is that gel-derived glasses have a fine-scale porosity. This is particularly evident with the acid-catalysed process. The pores are usually in the micropore (diameters below 2 nm) or mesopore (diameters greater than 2 nm but less than 50 nm) ranges. These classifications are the official IUPAC classifications [6].

The structure of a gel is established at the time of gelation. The size of the sol particles and the cross-linking within the particles (i.e. density) depend upon the pH and R ratio ($R = [H_2O]/[Si(OR)]$, molar ratio). The physical characteristics of the gel network depend greatly upon the size of particles (extent of cross-linking prior to gelation). The variables of major importance are temperature, nature and concentration of electrolyte (acid, base), nature of the solvent and type of alkoxide precursor.

The gel evolves from a sol, where the small particles that formed from the hydrolysis of an alkoxide precursor are weakly interacting with each other (Figure 3.4). The gel forms as condensation reactions occur between the particles, causing them to bond together. Bridging oxygen bonds link the particles together, forming a continuous network (Figure 3.5).

Primary particles of about 2 nm diameter agglomerate to form secondary particles of 5–100 nm diameter [7]. At low pH, O–Si–O bonds are formed and the condensation process resembles classical polycondensation, resulting in a three-dimensional silica network and depolymerisation is unlikely [3]. Gelation occurs when the secondary

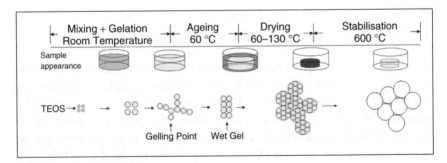

Figure 3.5 Schematic showing the evolution of a nanoporous glass network through the acid-catalysed sol-gel process. (Adapted with permission from [7] Copyright (2009) Royal Society of Chemistry.)

particles are linked to each other (Figure 3.5). The gelation continues during ageing. Figure 3.5 demonstrates that gel forms by particles coalescing and then bonding together. Spaces therefore form between the particles as they come together, as packing of spheres always leaves interstitial gaps between the spheres. When the gel is formed and is still wet, the by-products of hydrolysis and condensation remain in the pores and are called 'pore liquor'. During drying, the pore liquor evaporates, leaving behind the nanoporosity (Figure 3.1). Pore size is related to particle size, which in turn depends on reaction conditions and glass composition. The 'surfaces' of the pores are covered in Si–OH groups after drying following the evaporation of water. Thermal processing can be used to drive off the –OH groups and densify the silica network, which also reduces the pore size. However, for bioactive compositions, the thermal processing temperature is usually minimised to maintain nanoporosity.

Silica gels are mostly linear structures when formed under acidic conditions owing to the low degree of cross-linking due to steric crowding. In contrast, under basic conditions, the distribution of polysilicate species is broad and characteristic of branched polymers with a high degree of cross-linking [2]. Fast hydrolysis and slow condensation favour the formation of linear polymers. Slow hydrolysis and fast condensation result in larger, bulkier and more ramified networks [2]. Larger particles are anticipated when alcohol is used as a solvent (lower rate of depolymerisation). Removing the pore liquor from the nanoporosity without cracking the glass can be a challenge though.

3.6 MAKING SOL-GEL MONOLITHS

The greatest challenge in the sol-gel process is the production of crack-free monoliths. Discs, rods or blocks with diameters in excess of 1 cm are generally difficult to produce by the sol-gel process. This is particularly difficult if the glass has network modifiers, that is, if the composition is anything other than silica.

The primary problem that had to be overcome from the early days of the sol-gel process was cracking during drying. The cracking is due to two reasons: the large shrinkage that occurs during drying; and the evaporation of the liquid by-products of the condensation reaction. When pore liquids are removed from the gels, the vapour must travel from within the gel to the gel surface via the interconnected pore network. This can cause capillary stresses within the pore network and therefore cracking. For small cross-sections, such as in powders, coatings or fibres, drying stresses are small, because the path of evaporation is short and the stresses can be accommodated by the material. For monolithic objects greater than about 1 cm in diameter, the path from the centre of the monolith to the surface is tortuous and the drying stresses can introduce catastrophic fracture. For gels produced by the hydrolysis and condensation route, drying has to be carried out with low heating rates. Increasing pore size and obtaining pores with a narrow distribution reduces tortuosity.

The difference between the modern development of sol-gel derived materials, and the classical work of Ebelman is that now drying of the monolithic silica optics can be achieved in days rather than years.

3.7 MAKING PARTICLES

Particles can be made by crushing and grinding monoliths made by the acid-catalysed sol-gel process. In fact, when making powders, the drying process is less critical, because cracks are welcome as they facilitate grinding. Grinding sol-gel glasses is inherently easier than grinding melt-derived glasses because the sol-gel glass contains nanoporosity.

A more elegant method for making small particles is what has become known as the Stöber process [8]. The Stöber process uses sol-gel processing to produce spherical nanoparticles (or a least sub-micrometre particles). The process is based on colloidal processing. Using ammonium hydroxide as the catalyst rather than using acid tips the pH above the isoelectric point of soluble silica (silicic acid) [3]. The pH causes repulsion between newly formed silica particles and causes polycondensation to be terminated. Therefore, after primary particles form due to hydrolysis, some condensation occurs to form spherical secondary particles, but bonds do not form between the particles, so the secondary particles remain as particles (Figure 3.1b). The final size of the spherical silica powder can be controlled by several process variables: pH, the types of silicon alkoxide (methyl, ethyl, pentyl, esters, etc.) and alcohol (methyl, ethyl, butyl and pentyl) mixture used, and reactant temperature. In biomedical applications, small silica spheres have potential for cell labelling and drug delivery (see Figure 4 in colour section). This is because, if they are small enough to enter a cell and do not cause the cell to change behaviour, they can be used to carry therapeutic agents, for example small drug molecules.

Of particular benefit are small particles that contain nanopores. Drugs and growth factors can then be loaded into the particles and the payload can be delivered into cells by the particles. There is now a very large research area investigating mesoporous silica particles as drug delivery agents [9]. Mesoporous silica particles are also being designed to kill cancerous tumours [10]. The idea is that the particles, containing payloads that can kill tumour cells, are injected into the patient and the particles arrive at the tumour, using blood transport, where they are taken up by the cells. When the particles are inside the cells, they release their deadly payload. The aim is that the particles will target tumours so that they are not taken up by other (healthy) cells. This requires careful design of functional groups on the surface of the particles, so that they are attracted to the correct cells, and the use of other molecules that block the premature release of the payload before the particles reach the tumour.

For optimal control of drug loading, the porosity of the particles is often made to have an order to it (Figure 3.6). Ordered pores are made by adding a surfactant template to the sol. Surfactants, or 'surface-active agents', are short molecules with one end that is hydrophilic and the other hydrophobic. They are usually used to reduce the surface tension of a liquid, for example in detergents or to disperse particles in solution. When someone uses soap to wash their hands, bubbles form as air is entrapped in the water and the bubbles are stabilised by the surfactant. The hydrophilic end of the surfactant molecule attaches to water and the other end is in the air. Surfactant templates work slightly

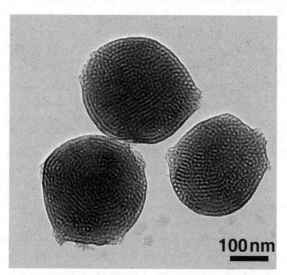

100 nm

Figure 3.6 Transmission electron microscope image of a silica particle containing ordered mesopores. (Image provided courtesy of Lijun Ji, College of Chemistry and Chemical Engineering, Yangzhou University, China.)

differently. The surfactant concentration is increased to the point of saturation, called the critical micelle concentration (CMC). At the CMC point, the surfactant molecules interact with each other, forming a sphere. The spheres are all similar size and now do not interact with the solution. However, the micelles pack themselves automatically (self-assemble) into hexagonal or cubic arrays and can used as templates, so the sol flows around them and, after gelation, the micelle spheres are removed by thermal processing similar to that used in the conventional sol-gel process.

3.8 SOL-GEL DERIVED BIOACTIVE GLASSES

Chapter 2 describes bioactive glasses. They are important materials because they bond to bone and dissolve safely in the body and stimulate bone repair. The original bioactive glass, Bioglass®, was melt-quenched and was a four-component system containing silica (SiO_2), soda (Na_2O), lime (CaO) and phosphate (P_2O_5) [11]. The soda was only in the composition to reduce the melting point of the glass, but the release of sodium into the body can be detrimental to cells. Bioactive sol-gel glasses are usually synthesised via the acid-catalysed route. They provide the benefit that glasses can be bioactive with much more simple compositions. Silica, lime, phosphate (58S, 60 mol% SiO_2, 36 mol% CaO, 4 mol% P_2O_5) compositions have recently gained approval for implantation into bone defects in the USA when mixed with NovaBone®, but glasses with 70 mol% SiO_2 and 30 mol% CaO (70S30C) are also highly bioactive [12]. The mechanism for bone bonding is similar to that of the melt-derived glasses (Chapter 2), with cation exchange with the body fluid (Ca^{2+} from the glass and H^+ from the fluid) on implantation followed by nucleation of a hydroxycarbonate apatite (HCA) layer, which forms the bond with bone.

Other active ions, such as silver and strontium, can also be incorporated into the sol-gel process. Silver has an impact in wound healing applications, because small amounts of silver are known to kill bacteria without harming healthy human cells. Silver can be released from the glass at the rate of degradation of the glass, and therefore sol-gel glasses can act as controlled delivery devices for silver [13]. Strontium ions have been shown to be beneficial to patients suffering from osteoporosis, as the ions inhibit osteoclast activity. Osteoclasts are the cells that remove bone in the natural remodelling process, but for osteoporosis sufferers the osteoclasts are much more active than the osteoblasts that are responsible for producing new bone, resulting in loss of bone density. Strontium incorporation in sol-gel glasses is an effective way to deliver a steady supply of strontium ions to a bone defect site [14].

A further advantage is that melt-quenched glasses cannot be bioactive if they have a silica content in excess of 60 mol%. For sol-gel glasses this is extended to 90 mol% [15]. The reason for this is the nanoporous network and the connectivity of the silica network. Owing to the wet nature of the process and the way the nanoparticles assemble, nanopores are left behind and the silica network is disrupted by protons (H^+ ions). The protons act as additional network modifiers. This means that the 70S30C composition is not only 70 mol% SiO_2 and 30 mol% CaO, but it also contains OH groups. Thus the network connectivity of the glass is lower than it would be for a melt-derived glass. The amount of OH groups decreases as the thermal processing temperature increases and therefore the real composition of the glass changes with thermal processing. The nanopores also cause the sol-gel glasses to have a surface area two orders of magnitude higher than that for similar melt-quenched compositions, which increases the degradation rate and therefore the rate of formation of the HCA layer that forms a bond to the apatite in bone. Tailoring the nanoporosity also provides an easy route to controlling the degradation rate of the glasses.

The addition of calcium to a silica sol-gel glass reduces the network connectivity of the glass and also increases the nanopore size. Calcium nitrate is the most common precursor for introducing calcium into a silica glass, as it is very soluble in the sol. However, calcium does not automatically incorporate into the silica network at room temperature. The calcium only enters the network when the temperature reaches 400 °C. This has an impact on low-temperature synthesis, such as that used for hybrid synthesis (Chapter 10).

A great benefit of the sol-gel process is that the process can be modified to produce porous scaffolds for bone repair or tissue engineering applications, either by foaming the sol (Figure 3.7), which produces foam-like scaffolds with pore structures similar to porous bone [16], or by feeding the sol into an electrospinning system to create fine fibrous membranes (Figure 3.8). Chapter 12 gives further details on these methods.

It is not only silica glasses that can be produced by the sol-gel process. Amorphous titania [17] and phosphate glasses [18] can also be produced by the sol-gel process. However, the titania is usually limited to thin films, and sol-gel phosphate glasses have, as yet, been very (too) soluble and fragile.

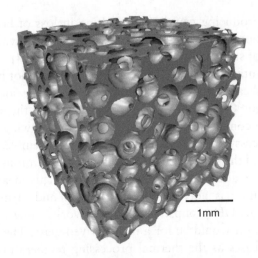

Figure 3.7 X-ray microtomography (micro-CT) image of a bioactive glass sol-gel foam.

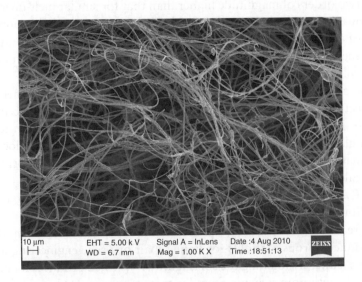

Figure 3.8 Sol-gel glass fibres made by electrospinning.

3.9 SUMMARY

The sol-gel process is a low-temperature process that utilises polymeric reactions to create an inorganic network (gel), which can be dried to form a glass. The glass is thought to be more homogeneous compared to melt-quenched glasses. Other differences include the fact that gel glasses have an inherent nanoporosity, which means their degradation rate can be controlled. Bioactive sol-gel derived glasses therefore have higher degradation rates and higher bioactivity than their melt-quenched counterparts. The process is flexible to allow processing into spherical particles, foams or fibres.

REFERENCES

[1] Hench, L.L. and West, J.K. (1990) The sol-gel process. *Chemical Reviews*, **90**, 33–72.

[2] Brinker, J. and Scherer, G.W. (1990) *Sol-Gel Science: The Physics and Chemistry of Sol-Gel Processing*. Boston, MA: Academic Press.

[3] Iler, R.K. (1979) *The Chemistry of Silica: Solubility, Polymerization, Colloid and Surface Properties and Biochemistry of Silica*. New York: Wiley-Interscience.

[4] Ebelmen, M. (1846) Recherches sur les combinaisons des acides borique et silicique avec les ethers. *Annales de Chimie et de Physique*, **116**, 129–166.

[5] Prassas, M., Phalippou, J. and Zarzycki, J. (1984) Synthesis of monolithic silica-gels by hypercritical solvent extraction. *Journal of Materials Science*, **19**, 1656–1665.

[6] Sing, K.S.W., Everett, D.H., Haul, R.A.W. *et al.* (1985) Reporting physisorption data for gas–solid systems with special reference to the determination of surface-area and porosity (recommendations 1984). *Pure and Applied Chemistry*, **57**, 603–619.

[7] Lin, S., Ionescu, C., Pike, K.J. *et al.* (2009) Nanostructure evolution and calcium distribution in sol-gel derived bioactive glass. *Journal of Materials Chemistry*, **19**, 1276–1282.

[8] Stöber, W. (1968) Controlled growth of monodisperse silica spheres in micron size range. *Journal of Colloid and Interface Science*, **26**, 62.

[9] Vallet-Regi, M., Balas, F. and Arcos, D. (2007) Mesoporous materials for drug delivery. *Angewandte Chemie – International Edition*, **46**, 7548–7558.

[10] Ashley, C.E., Carnes, E.C., Phillips, G.K. *et al.* (2011) The targeted delivery of multicomponent cargos to cancer cells by nanoporous particle-supported lipid bilayers. *Nature Materials*, **10**, 389–397.

[11] Hench, L.L. and Polak, J.M. (2002) Third-generation biomedical materials. *Science*, **295**, 1014–1017.

[12] Saravanapavan, P., Jones, J.R., Pryce, R.S. and Hench, L.L. (2003) Bioactivity of gel-glass powders in the $CaO-SiO_2$ system: a comparison with ternary $(CaO-P_2O_5-SiO_2)$ and quaternary glasses $(SiO_2-CaO-P_2O_5-Na_2O)$. *Journal of Biomedical Materials Research Part A*, **66A**, 110–119.

[13] Bellantone, M., Williams, H.D. and Hench, L.L. (2002) Broad-spectrum bactericidal activity of Ag_2O–doped bioactive glass. *Antimicrobial Agents and Chemotherapy*, **46**, 1940–1945.

[14] Isaac, J., Nohra, J., Lao, J. *et al.* (2011) Effects of strontium-doped bioactive glass on the differentiation of cultured osteogenic cells. *European Cells and Materials*, **21**, 130–143.

[15] Pereira, M.M., Clark, A.E. and Hench, L.L. (1995) Effect of texture on the rate of hydroxyapatite formation on gel-silica surface. *Journal of the American Ceramic Society*, **78**, 2463–2468.

[16] Jones, J.R., Ehrenfried, L.M. and Hench, L.L. (2006) Optimising bioactive glass scaffolds for bone tissue engineering. *Biomaterials*, **27**, 964–973.
[17] Kokubo, T., Kim, H.M. and Kawashita, M. (2003) Novel bioactive materials with different mechanical properties. *Biomaterials*, **24**, 2161–2175.
[18] Carta, D., Knowles, J.C., Guerry, P. *et al.* (2009) Sol-gel synthesis and structural characterisation of P_2O_5–B_2O_3–Na_2O glasses for biomedical applications. *Journal of Materials Chemistry*, **19**, 150–158.

4

Phosphate Glasses

Delia S. Brauer
Otto Schott Institute, Friedrich Schiller University Jena, Jena, Germany

4.1 INTRODUCTION

Phosphate glasses, as the name suggests, consist of phosphate (rather than silicate or borosilicate) as the glass former. For example, calcium phosphate glasses can have a composition similar to the mineral phase of bone, which is a biological calcium phosphate (a hydroxycarbonate apatite), which makes them an interesting material as a synthetic bone graft [1]. In fact, phosphate glasses can dissolve completely in aqueous solution, giving ionic species commonly found in the human body [1], which offers various possibilities for their application as degradable implant materials. Therapeutic ions such as strontium, zinc or fluoride can be incorporated into the glass to be released upon degradation for stimulation of bone growth and wound healing and for prevention of infections, which can make phosphate glasses a very versatile biomaterial for tissue regeneration.

Bio-Glasses: An Introduction, First Edition. Edited by Julian R. Jones and Alexis G. Clare.
© 2012 John Wiley & Sons, Ltd. Published 2012 by John Wiley & Sons, Ltd.

4.2 MAKING PHOSPHATE GLASSES

Phosphate glasses are commonly produced by mixing the glass precursors (powders of phosphates, oxides and carbonates; sometimes phosphoric acid is used as a phosphate source) and melting at high temperatures in a platinum crucible. Alumina crucibles should be avoided in order to prevent aluminium contamination of the glasses, as aluminium is a neurotoxin and negatively affects bone mineralisation. The melting temperatures of phosphate glasses tend to be lower than those of bioactive silicate glasses, and are usually between 800 and 1300 °C, depending on the glass composition. After the material has melted completely, the melt is cooled by either quenching quickly between metal plates or by pouring into water. In order to obtain bulk glass (monoliths), the hot melt is poured into preheated graphite or metal moulds, placed into a furnace preheated to the transition temperature of the glass and annealed, that is, allowed to cool down to room temperature slowly in order to reduce stresses in the glass.

An alternative route for making glasses is by a sol-gel route. However, sol-gel processing of phosphate glasses is relatively new [2], the glasses produced seem to be more fragile than bioactive silicate sol-gel glasses (Chapter 3) and they are very soluble, possibly too soluble at the time of writing for most biomedical applications.

4.3 PHOSPHATE GLASS STRUCTURE

Like silicate glasses, phosphate glass structure does not show any long-range order or significant symmetry of atomic arrangement, but they do have short-range order. The glass-forming component in phosphate glasses is P_2O_5, and the basic unit in the phosphate glass structure is the orthophosphate (PO_4^{3-}) tetrahedron (Figure 4.1), which is a phosphorus atom surrounded by four oxygen atoms. As one of the oxygen atoms is connected to the phosphorus atom by a double bond, only the three other oxygen atoms can act as 'bridges' to other orthophosphate tetrahedra. By forming such bridges, individual orthophosphate tetrahedra can be

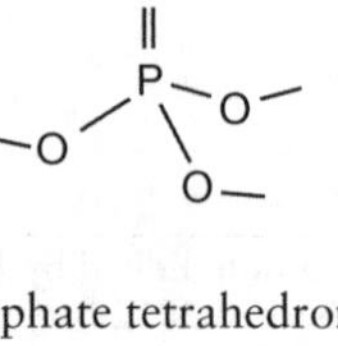

Figure 4.1 Basic phosphate tetrahedron in glass structures.

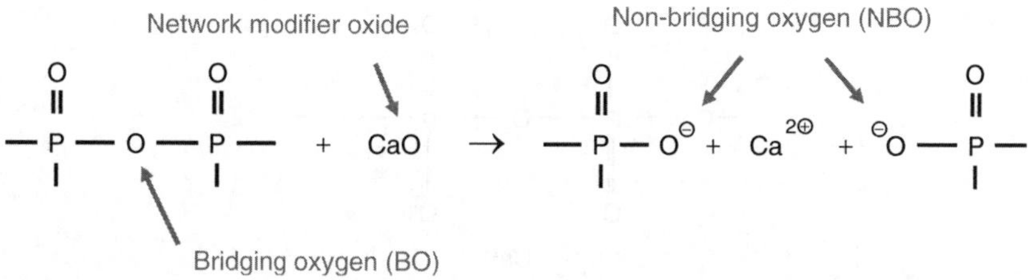

Network modifier oxide   Non-bridging oxygen (NBO)

$$\begin{array}{c} O \\ \parallel \\ -P-O-P- \\ \mid \qquad\quad \mid \end{array} + CaO \rightarrow \begin{array}{c} O \\ \parallel \\ -P-O^\ominus + Ca^{2\oplus} + {}^\ominus O-P- \\ \mid \qquad\qquad\qquad\qquad\quad \mid \end{array}$$

Bridging oxygen (BO)

Figure 4.2 Schematic of cleavage of P–O–P linkages by network modifier oxides, turning bridging oxygens (BOs) into non-bridging oxygens (NBOs).

linked to each other by covalent POP bonds. The oxygens in these POP linkages are commonly called bridging oxygens (BOs). Orthophosphate tetrahedra can thereby be arranged to form chains, rings or branching networks.

P_2O_5 is called a network former because it is possible to make a glass using P_2O_5 only. Pure P_2O_5 glass can be produced under certain conditions, but, as it is very reactive and hygroscopic, it is of scientific interest only. The durability can be increased, however, by adding other components, called network modifiers, such as calcium oxide (CaO).

The addition of network modifier oxides such as Na_2O and CaO to the glass results in the cleavage of POP linkages and the creation of non-bridging oxygens (NBOs) in the glass, as shown in Figure 4.2. The resulting glass structure is disrupted, containing not only covalent POP bonds but also ionic cross-linkages between NBOs, and this process is called depolymerisation of the network (Figure 4.3).

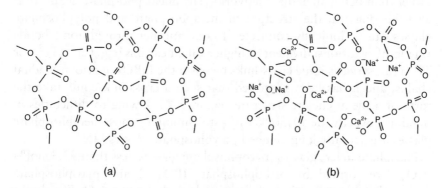

(a)   (b)

Figure 4.3 Schematic of phosphate glass structure: (a) P_2O_5 glass consisting of a phosphate network and (b) phosphate glass after addition of modifier cations and cleavage of P–O–P bonds, turning bridging oxygens (BOs) into non-bridging oxygens (NBOs).

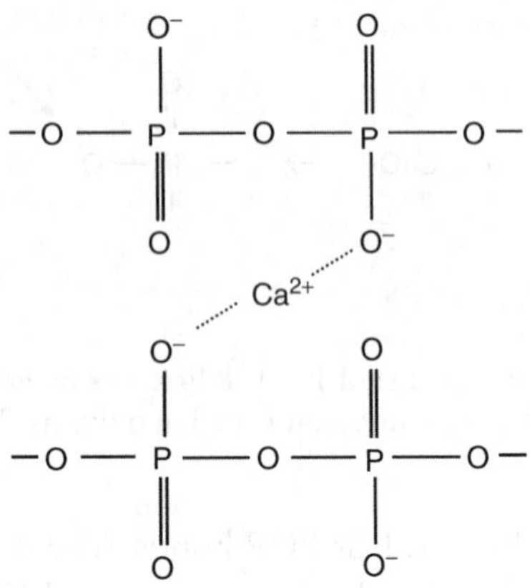

Figure 4.4 Simplified chelate structure of ionic cross-links between phosphate chains.

Corresponding to their structure, phosphate glasses can be divided into different groups [3]. Ultraphosphate glasses contain more than 50 mol% P_2O_5, that is, less than 50 mol% of network modifier oxides, and accordingly they consist of two- to three-dimensional phosphate networks. Polyphosphate glasses contain less than 50 mol% P_2O_5, and they are built of phosphate chains and rings, with the chain length decreasing with decreasing phosphate content. In between these are metaphosphate glasses that contain 50 mol% P_2O_5. Their structure also consists of chains of infinite length or rings, and is therefore better described as consisting of entangled 'molecular' or 'polymeric' chains, rather than being an actual 'network'. The linear phosphate chains (but also the rings) in the structure of metaphosphate and polyphosphate glasses can be ionically connected to one another through ionic bonds via modifier cations; divalent or higher valent cations (e.g. Ca^{2+}, Fe^{3+} or Ti^{4+}) can serve as ionic cross-links between the NBOs of two individual chains, and it has been suggested that such a cross-link could take the form of a metal chelate structure (Figure 4.4). Owing to difficulties in obtaining exactly the 50 mol% P_2O_5 stoichiometry, most metaphosphate glasses are actually long-chained polyphosphate glasses [4].

Phosphate invert glasses (pyrophosphate glasses, less than 33.3 mol% P_2O_5) are formed by orthophosphate (PO_4^{3-}) and pyrophosphate groups ($P_2O_7^{4-}$; phosphate dimers) exclusively (Figure 4.5). For glasses with lower phosphate contents, isolated orthophosphate groups are present. In these cases, the glassy state is caused neither by a relatively stiff network nor by entangled chains but by the interaction of

PO$_4^{3-}$ tetrahedron

P$_2$O$_7^{4-}$
pyrosphosphate
group

modifier cations
(Ca^{2+}, Na$^+$...)

Figure 4.5 Schematic of invert glass structure.

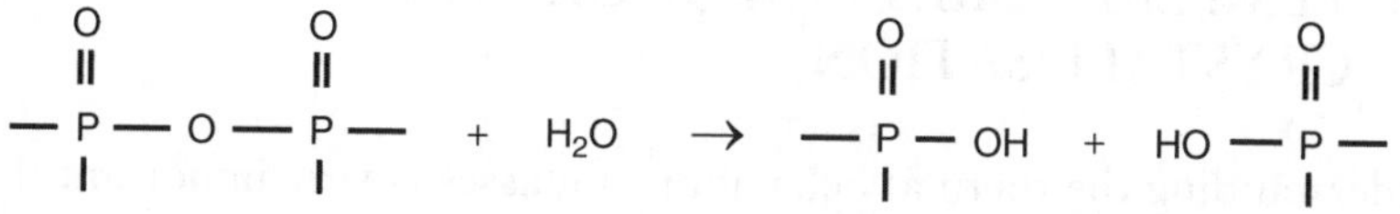

$$\underset{|}{\overset{O}{\overset{\|}{-P}}} - O - \underset{|}{\overset{O}{\overset{\|}{P}}} - \quad + \quad H_2O \quad \rightarrow \quad \underset{|}{\overset{O}{\overset{\|}{-P}}} - OH \quad + \quad HO - \underset{|}{\overset{O}{\overset{\|}{P}}} -$$

Figure 4.6 Schematic of structural water causing cleavage of P–O–P linkages.

cations and phosphate groups, which is why they are often referred to as *invert* glasses.

As described above, the structure of phosphate glasses depends on their phosphate (P$_2$O$_5$) content and on the content of network modifier oxides, such as CaO or Na$_2$O. However, water can also disrupt the phosphate network, thereby acting like a network modifier. Water, for example from the atmosphere, can disrupt POP linkages, creating POH groups instead (Figure 4.6). This reaction occurs readily for vitreous P$_2$O$_5$ and most ultraphosphate glasses; however, it can also affect actual chain lengths in metaphosphate glasses, resulting in shorter phosphate chains than predicted based on the composition. With decreasing chain length, phosphate glasses are less prone to water attack and subsequent chain scission, and are therefore more stable when exposed to atmospheric humidity.

Glass properties, such as crystallisation tendency, mechanical properties or stability against hydrolytic attack, depend not only on the phosphate content (and the phosphate glass structure) but also on the size and the charge (i.e. the charge-to-size ratio) of the network modifier cation, which determines how strong the ionic cross-links between two NBOs are: an increase in charge-to-size ratio results in stronger cross-linking. Therefore, cross-linking would be expected to increase in the order $Na^+ \rightarrow Ca^{2+} \rightarrow Fe^{3+} \rightarrow Ti^{4+}$; similarly for cations of the same charge but with decreasing ionic radius, as for the series $Ba^{2+} \rightarrow Sr^{2+} \rightarrow Ca^{2+} \rightarrow Mg^{2+}$.

Besides network formers and network modifiers, there is a third class of glass components, called intermediate oxides [5]. Intermediate oxides can switch their role, either acting like a network modifier, that is, disrupting P–O–P linkages and creating NBOs, or acting more like a network former, entering the glass network (i.e. the phosphate network), creating P–O–M–O–P bonds (M = intermediate metal atom) with a more covalent character. Magnesium (Mg), iron (Fe), aluminium (Al), zinc (Zn) and titanium (Ti), among others, are typical intermediates.

4.4 TEMPERATURE BEHAVIOUR AND CRYSTALLISATION

Understanding the thermal behaviour of glasses is very important if the glasses are to be manipulated or processed into complex shapes. Besides their amorphous structure, their temperature behaviour is another characteristic of glassy materials. While crystalline materials show a distinct melting point, where they transform from a solid material to a liquid, glasses show a temperature range of glass transition or transformation, in which they soften (i.e. their viscosity decreases) gradually. The glass transition temperature (T_g) is usually defined as the onset of this transformation temperature range. The value of T_g can be determined using methods such as differential thermal analysis (DTA) or differential scanning calorimetry (DSC), and in both methods T_g appears as a shift in the baseline (Figure 4.7).

Glass transition and crystallisation temperatures decrease with increasing phosphate content (i.e. from invert to ultraphosphate glasses), as shown for binary polyphosphate glasses in the system P_2O_5–MgO (Figure 4.8) [6]. This trend can be explained by decreasing ionic cross-linking for less disrupted glasses, resulting in lower T_g.

In general, the more disrupted the phosphate structure is (i.e. the smaller the phosphate units), the more easily the glass crystallises. The tendency to crystallise is closely connected to the viscosity of the melt, with lower viscosities allowing for the components to arrange into an ordered crystalline structure more easily. This means that phosphate

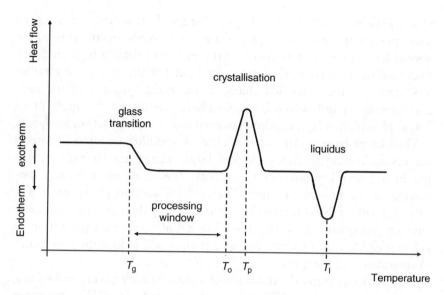

Figure 4.7 Schematic of DSC trace of a glass showing glass transition temperature (T_g), crystallisation onset temperature (T_o), crystallisation peak temperature (T_p) as well as liquidus temperature (T_l).

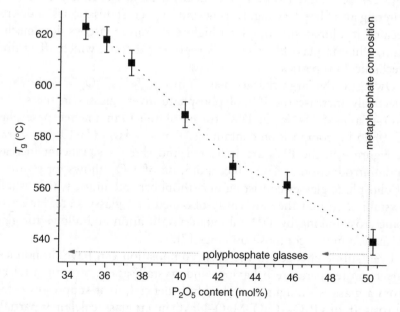

Figure 4.8 Glass transition temperature (T_g) of binary P_2O_5–MgO glasses versus nominal P_2O_5 content. (Adapted with permission from [6]. Copyright (2003) Elsevier Ltd.)

invert glasses, which have a highly disrupted structure consisting of short phosphate units (orthophosphate and pyrophosphate groups, see above) have a very low viscosity of the melt and show a high tendency to crystallise. Phosphate glasses consisting of a chain structure crystallise less easily, as the entangled chains increase the viscosity of the melt and impede crystallisation. Ultraphosphate glasses, which consist of an actual phosphate network, show an even lower crystallisation tendency.

The charge-to-size ratio of the network modifier cation also affects the crystallisation tendency, with a larger charge-to-size ratio result-ing in a lower crystallisation tendency due to stronger ionic cross-linking of the phosphate units. Therefore, a sodium phosphate glass, $xP_2O_5-(100-x)Na_2O$, crystallises more readily than the corresponding calcium phosphate glass, $xP_2O_5-(100-x)CaO$, while incorporation of calcium oxide into a sodium phosphate glass will reduce the crystallisa-tion tendency of the glass.

In order to quantify the tendency of a glass to undergo crystallisation, its processing window (PW) can be determined. The PW (sometimes referred to as the sintering window) is the temperature range between glass transition temperature (T_g) and the onset of crystallisation (T_o, Figure 4.7). A large PW ($>100\,K$, ideally around $200\,K$) is preferred for sintering and fibre drawing. During sintering, crystallisation is undesired because it inhibits sintering by a high-temperature viscous flow mecha-nism, while in glass fibres crystals represent defects, which affect fibre mechanical properties.

Owing to the large charge-to-size ratio of Ti^{4+}, TiO_2 was shown to effectively increase the PW of phosphate invert glasses in the system $P_2O_5-CaO-MgO-Na_2O$. DSC traces of the titanium-free base glass (Ti-0) and a composition containing 10 mol% TiO_2 (Ti-10) are shown in Figure 4.9; the PWs are indicated and show a significant increase upon introduction of TiO_2. Incorporation of TiO_2 allows for synthesis of phosphate glasses consisting of orthophosphate units, which usually crystallise, so that they are easily obtained in a glassy state. However, ionic cross-linking by Ti^{4+} inhibits crystallisation and allows for glass formation in this composition range [7].

Even subtle changes to the glass composition can have pronounced effects on temperature behaviour and crystallisation. Strontium release from a glass is thought to be highly beneficial, as it suppresses osteo-porosis. If, in a $P_2O_5-CaO-MgO-Na_2O$ invert glass, calcium is partially replaced by strontium, the PW decreases from about $145\,K$ in the all-calcium glass to $125\,K$ in the all-strontium glass (Figure 4.10). Although the ionic radius (and subsequently the charge-to-size ratio)

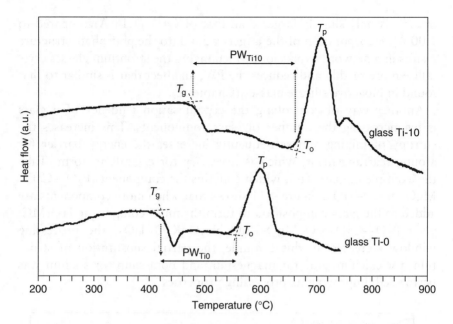

Figure 4.9 DSC traces of two phosphate invert glasses showing glass transition temperature (T_g), crystallisation onset temperature (T_o) and crystallisation peak temperature (T_p) and the processing window (PW). Glass Ti-10 shows a larger processing window than glass Ti-0 due to strong ionic cross-linking by Ti^{4+}.

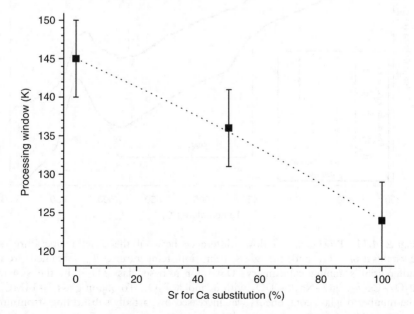

Figure 4.10 Processing window (temperature difference between crystallisation onset, T_o, and glass transition, T_g) versus Sr for Ca substitution in the glass.

of Sr^{2+} is only slightly larger than that of Ca^{2+} (1.18 Å compared to
1.00 Å), incorporation of the larger cation into the phosphate structure
results in a network expansion, which makes the strontium glasses crystallise more readily and reduces the PW, an effect that is similar to that
found in bioactive silicate glasses (Chapter 2).

Another way of controlling the crystallisation tendency of a glass
is by increasing the number of glass components. This increases the
entropy of mixing and subsequently increases the energy barrier for
atomic rearrangement, which is necessary for crystals to form. Take
the example of glass Ti-3, which contains the components P_2O_5–CaO–
MgO–Na_2O–TiO_2. Figure 4.11 shows that when more components are
added to the glass composition to form the multi-component Ti-3(MC)
glass P_2O_5–CaO–SrO–MgO–ZnO–Na_2O–K_2O–TiO_2, the processing
window increases. In this example, the partial substitution of strontium for calcium, zinc for magnesium and potassium for sodium was
performed, improving its processing properties.

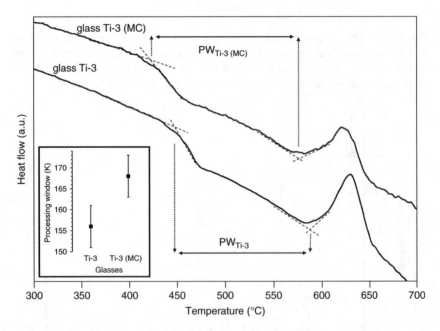

Figure 4.11 Processing window (difference between the onset temperature of
crystallisation, T_o, and the glass transition temperature, T_g, obtained from
differential scanning calorimetry, DSC) for a base glass (Ti-3) in the system
P_2O_5–CaO–MgO–Na_2O–TiO_2 with 37 mol% P_2O_5. To obtain glass Ti-3 (MC),
the number of glass components was increased by partially substituting strontium
for calcium, zinc for magnesium and potassium for sodium.

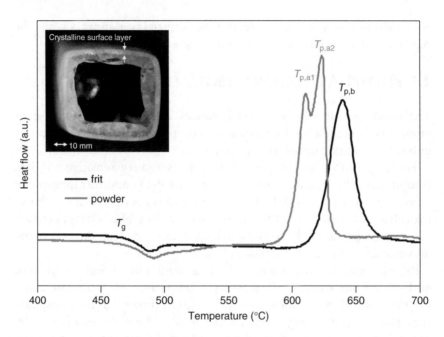

Figure 4.12 DSC traces of frit (particle size about 1 mm) and fine powder of a phosphate invert glass in the system P_2O_5–CaO–MgO–Na_2O–TiO_2 with 34.8 mol% P_2O_5. Traces show glass transition (T_g) and crystallisation temperatures ($T_{p,a1}$, $T_{p,a2}$ and $T_{p,b}$). Inset shows surface crystallisation on glass monolith after heat treatment for 30 minutes at 540 °C.

When comparing the DSC traces of frit (i.e. particles obtained by quenching the glass melt in water) and milled fine powder of a phosphate invert glass (34.8 mol% P_2O_5) in Figure 4.12, two main differences are obvious: the crystallisation peak of the fine powder appears at a lower temperature than that of frit; and it actually consists of two overlapping peaks, compared to only one visible peak for glass frit. The reason for the double peak is that the glass crystallises into two different phases (identified as calcium pyrophosphate, $Ca_2P_2O_7$, and a mixed calcium magnesium pyrophosphate, $CaMgP_2O_7$), which are resolved as separate peaks in the DSC trace of fine glass powder but not for glass frit. The decrease in crystallisation temperature with decreasing particle size (from coarse frit to fine powder), on the other hand, often suggests a surface crystallisation mechanism. Surface crystallisation means that, upon heat treatment, the crystals start growing from the outer surface of the glass towards the centre, rather than throughout the bulk of the material as in volume crystallisation. Indeed, if a glass monolith of the same composition is heat-treated, a crystalline surface layer appears while the centre of the monolith is free of crystals (see inset in Figure 4.12).

4.5 PHOSPHATE GLASS DISSOLUTION

The main reason why phosphate glasses are of interest for use as biomaterials is their ability to dissolve completely in aqueous solutions into safe non-toxic dissolution products.

The dissolution rate of phosphate glasses is very sensitive to glass composition [8]. For dissolution of vitreous P_2O_5 and ultraphosphate glasses, cleavage of P–O–P bonds is necessary. As mentioned above, hydrolysis of vitreous P_2O_5 occurs readily, but also ultraphosphate glasses are prone to hydrolysis, which results in degradation in the presence of atmospheric humidity.

By contrast, for dissolution of meta- and polyphosphate glasses, no P–O–P bond cleavage is necessary. Comparison of phosphate chain lengths in the glass and those found in solution shows that the phosphate chains stay intact during glass dissolution [8]. These glasses dissolve by hydration of entire phosphate chains and subsequent chain disentanglement and dissolution. Hydrolysis of polymeric phosphate chains occurs in solution after the initial hydration and dissolution, but at a much slower rate.

In general, solubility decreases with decreasing P_2O_5 content. Figure 4.13 shows glass dissolution (presented as dissolved phosphate relative to total phosphate in the glass) as a function of phosphate content in the glass, and a general trend confirms an increase in solubility with increasing phosphate content, in the order invert glasses → polyphosphate glasses → metaphosphate glasses.

When phosphate glasses dissolve, they affect the pH of the surrounding solution differently to silicate glasses (Chapter 2). Metaphosphate glasses are known to give an acidic pH, while decreasing phosphate content increases the pH change towards neutral (Figure 4.14). The resulting pH, however, depends on several factors, such as the ratio of glass to liquid, fluid flow/exchange or the buffering capacity of the solution.

The pH of the surrounding solution also affects glass dissolution. In an acidic environment, phosphate glass dissolution increases dramatically [8], while a basic pH also increased glass dissolution (compared to pH 7) although at a slower rate. This shows that maintaining the pH in a neutral range is important if glass degradation needs to be controlled (e.g. for degradable implant materials). If a phosphate glass, such as a metaphosphate glass, lowers the local pH upon initial degradation, it

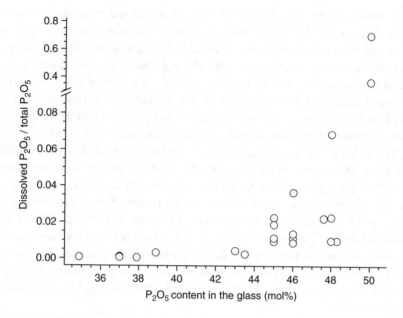

Figure 4.13 Dissolved P_2O_5 relative to total amount of P_2O_5 versus phosphate content in the glass.

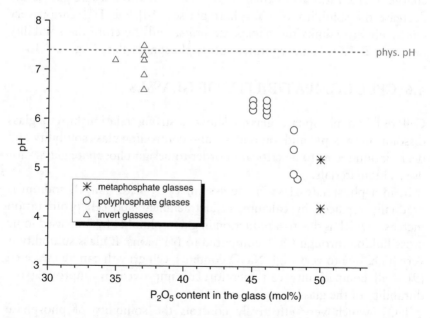

Figure 4.14 pH in physiological NaCl solution (9 g l^{-1}) versus phosphate content in the glass. The dashed line represents the physiological pH of 7.4.

will subsequently dissolve even faster, as its degradation rate increases with decreasing pH. This process is commonly known as autocatalysis. It is important to note that, while a buffered system can control pH changes to some extent, local pH changes near the glass surface can be quite dramatic (even in buffered systems such as *in vivo*), resulting in rapid degradation and failure of the implant. Fluid flow around an implant is therefore an important factor, as fluid flow would remove the cations from the implant surface, reducing the autocatalytic effect.

Apart from the phosphate content and the pH of the surrounding solution, the charge-to-size ratio of the modifier cation was also found to affect phosphate glass solubility. As described above, an increase in the charge-to-size ratio results in more effective cross-linking of the phosphate units, and the stronger this cross-linking, the lower the glass solubility. The solubility of phosphate glasses is therefore directly related to their composition. If, in a phosphate glass in the system P_2O_5–CaO–Na_2O, calcium is gradually replaced by sodium, that is, if the alkali content increases, the solubility increases and the durability decreases [9].

Glass solubility also increases with increasing ionic radius by constant charge, as in the series $Li^+ \rightarrow Na^+ \rightarrow K^+$ [8]. The effect of the charge-to-size ratio also explains why TiO_2 was found to significantly decrease the solubility of phosphate glasses [10]: the Ti^{4+} cation very effectively cross-links the phosphate units, and therefore the solubility decreases if TiO_2 is incorporated into a phosphate glass (Figure 4.15).

4.6 CELL COMPATIBILITY OF GLASSES

Cell tests on phosphate glasses show a strong relationship to glass dissolution rate, pH and ion release, and controlling glass solubility and degradation seems to be critical in order to design phosphate glasses for use as biomaterials.

If, in a phosphate glass in the system P_2O_5–CaO–Na_2O, sodium is gradually replaced by calcium, cell attachment and cell proliferation increase, which is due to a reduction in glass solubility by increased ionic cross-linking through Ca^{2+} compared to Na^+ ions. If glass solubility is very high, due to very high Na_2O contents, cell growth can be inhibited [9]. Cell compatibility can therefore be improved by improving the durability of the glasses.

TiO_2, which very effectively controls the solubility of phosphate glasses, can modulate cell adhesion and proliferation. Cells cultured on TiO_2-containing metaphosphate glasses show enhanced proliferation

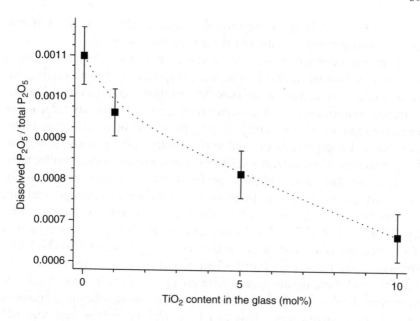

Figure 4.15 Dissolved P_2O_5 relative to total amount of P_2O_5 versus TiO_2 content in an invert phosphate glass of the composition P_2O_5–CaO–MgO–Na_2O (P_2O_5 content constant at 37 mol%).

and differentiation compared to a TiO_2-free control, an effect that can be explained by lower degradation rates and less pronounced pH changes for increasing TiO_2 content (glasses gave slightly acidic pH values, with a decreasing TiO_2 content corresponding to a decrease in pH) [9].

Replacing calcium with strontium in TiO_2-containing metaphosphate glasses did not have any marked effects on cell proliferation over up to seven days [11]. Although strontium is known to stimulate osteoblasts and to inhibit osteoclasts *in vitro* [12], and it is also the basis of a drug (Protelos®, Servier) for treating osteoporosis, the increase in solubility (due to the larger ionic radius of Sr^{2+} compared to Ca^{2+}) and the subsequent pH decrease apparently overcame the beneficial effects of strontium.

Zinc-containing metaphosphate glasses (P_2O_5–CaO–Na_2O–ZnO, with ZnO contents between 1 and 20 mol%) showed attachment of osteoblast-like cells, but cells maintained a round morphology and did not spread on the glasses, indicating that the cells were not happy on the surface [13]. These findings can be explained by potentially negative effects of zinc: while zinc was shown to have a stimulatory action on bone formation, higher than optimum levels of zinc were shown to be toxic [14]. Particularly for degradable biomaterials, the control of zinc content and subsequent zinc release is therefore essential.

In vitro cell culture analyses are a valuable tool for the assessment of a variety of biological and biochemical responses to biomaterials. This is the case, in particular, when specific chemical properties of individual material components, their concentrations, toxicity and stimulatory or negative regulatory effects on cell growth, are to be determined. However, some limitations associated with *in vitro* culture analyses need to be considered when interpreting the results. For example, results from cell tests on glass discs and those performed using glass extracts cannot be considered as equivalent. In addition, it is difficult to extrapolate from *in vitro* results to the *in vivo* behaviour of a material, as demonstrated by Navarro *et al.* [15], who showed that both a TiO_2-containing and a TiO_2-free polyphosphate glass in the system $P_2O_5–CaO–Na_2O–(TiO_2)$ showed good biocompatibility when tested in subcutaneous tissue in rabbits, and that, despite great differences in solubility and results of *in vitro* cell culture studies, the glasses showed no significant differences in their *in vivo* response. This can be explained by the fact that pH buffering by the physiological environment and the continuous circulation of body fluids were able to compensate for changes in chemical conditions near the implant.

4.7 PHOSPHATE GLASS FIBRES AND COMPOSITES

Glasses are commonly known to be brittle, and fabrication of glass fibres (Figure 4.16) is a way to produce a glassy material with improved mechanical properties. The excellent mechanical properties of glass fibres have already opened a broad field of applications, for example, fibre-reinforced plastics, a group of new materials with very low weight for highest mechanical and chemical stresses. The good mechanical properties of glass fibres make degradable phosphate glass fibres of interest in the field of biomaterials, for example, fabrication of bone fracture fixation materials such as glass-fibre-reinforced degradable polymers, but also as meshes or woven fibre constructs for tissue engineering or soft tissue replacement.

Conventionally, fibres are produced by drawing directly from the glass melt onto a rotating drum; ideal viscosities for fibre drawing lie between 10^3 and 10^6 Pa s, as melts with lower viscosity cannot easily be drawn into fibres. The fibre diameter can be adjusted by varying the drawing speed, with increasing speed resulting in thinner fibres. An alternative method involves making a preform (usually a glass rod), which will

Figure 4.16 Phosphate glass fibres in the system $P_2O_5-CaO-Na_2O-SiO_2$; fibre diameter is around 30 µm. (Image provided courtesy of C. Rüssel, Jena, Germany. Copyright (2012) Otto-Schott-Institut.)

be heated to temperatures above T_g and then drawn into fibres. This method is particularly interesting for producing fibres of highly disrupted polyphosphate or invert phosphate glasses, which show a high tendency to crystallise. Crystallisation of glass fibres can have deleterious effects on fibre production and mechanical properties, and it also influences glass solubility, degradation rates and cell response.

One of the major challenges during development of both conventional and degradable implant materials for fracture fixation is the mechanical compatibility between implants and bone. Metals and alloys have been used successfully for internal fixation; however, these implants do not degrade over time, and the rigid fixation from bone plating can cause stress protection atrophy resulting in loss of bone mass and osteoporosis. While the elastic modulus of cortical bone ranges from 17 to 26 GPa, common alloys have moduli ranging from 100 to 200 GPa. This large difference in stiffness can result in high stress concentrations as well as relative motion between the implant and bone upon loading. Degradable polymers, such as poly(lactic acid) (PLA), are currently used as sutures or degradable fracture fixation materials such as screws. Their main advantages over metallic implants are avoidance of a second operation to remove the fixation device and avoidance of stress shielding by a gradual load transfer from the degrading polymer to the regenerating bone. However, their lower stiffness (for PLA screws, around 3 GPa)

in comparison to bone may allow too much bone motion for satisfactory healing, and reinforcement therefore is essential. Polymeric
self-reinforced screws showed higher tensile and bending strength in
comparison with homogeneous polymeric screws; however, the elastic
moduli were still too low, resulting in bending of the screws, which
limited their use.

As an alternative, degradable polymers can be reinforced with phosphate glass fibres. Reinforcement by glass fibres is of interest as
the fibres maintain stability and mechanical properties during later
states of polymer degradation. Indeed, glass-fibre-reinforced composites
offer excellent strength and stiffness, and continuous phosphate glass-
fibre-reinforced degradable polymer composites can have mechanical
properties suitable for fixing cortical bone fractures [16, 17], with elastic
moduli between 15 and 40 GPa.

4.8 APPLICATIONS

Although the importance of phosphate glasses compared to bioactive
silicate glasses has been small so far, their solubility, which ranges
over several orders of magnitude and can be tailored by changing the
glass composition, makes them a promising class of materials in the
field of biomedical materials. By making glass fibres or sintered porous
scaffolds, substrates for a variety of applications can be produced. There
is great interest in the development of phosphate glasses, and particularly
degradable phosphate glass–polymer composites, for application in bone
fracture fixation and hard tissue regeneration. The aim is to develop an
implant material that slowly degrades in the body, to be replaced by
newly formed natural bone. Controlling the glass solubility is a key issue
here in order to match the resorption rate of the implant to the rate of
bone growth.

Also, for soft tissue applications such as engineering of ligament, muscle or cartilage regeneration, the use of phosphate glasses, for example
as fibres, meshes or in composites, would be of interest. It has further
been suggested that degradable phosphate glasses could potentially be
used for nerve regeneration and neural repair [1]. Despite a significant
increase in research output on phosphate glasses for biomedical applications, however, further research is necessary to optimise the materials
for *in vivo* applications.

Fluoride-releasing phosphate glass-based devices for prevention of
dental cavities are giving promising results for controlling caries in
children [18]. Low levels of fluoride are known to prevent enamel

demineralisation, enhance remineralisation and inhibit the metabolism of acidogenic bacteria such as mutans streptococci and lactobacilli [19]. Fluoride-releasing phosphate glass pellets are attached to molars, and while the glasses continuously degrade inside the oral cavity, they release fluoride, thereby allowing for low levels of fluoride to be maintained in the mouth.

The ability of phosphate glasses to release ions upon degradation can also be used in the development of materials for controlled release of therapeutically active ions, such as strontium-releasing glasses for treatment of osteoporosis, or for release of antibacterial ions such as copper, zinc or silver [9].

In addition to medical uses, there are other applications where the solubility of phosphate glasses is of interest; and phosphate glasses are used as fertilisers in agriculture and for treating trace-element deficiency in cattle [20]. In the latter case, glass boluses are placed in the animals' stomachs, where they give a sustained release of ions (e.g. cobalt, copper, selenium or zinc) over time.

4.9 SUMMARY

Phosphate glasses are soluble in aqueous solution, and can have a composition similar to mineralised tissue, which makes them of interest for use as degradable materials for bone regeneration, but also for use as materials for controlled release of ions in therapeutic concentrations. Glass solubility is closely related to its composition and structure, and controlling degradation and ion release are critical for successful applications of these materials. Composite materials based on degradable polymers such as PLA and phosphate glasses are a promising way of modulating the mechanical properties to match them to the surrounding tissue in the body.

REFERENCES

[1] Knowles, J.C. (2003) Phosphate based glasses for biomedical applications. *Journal of Materials Chemistry*, **13**, 2395–2401.
[2] Carta, D., Knowles, J.C., Guerry, P. *et al.* (2009) Sol-gel synthesis and structural characterisation of P_2O_5–B_2O_3–Na_2O glasses for biomedical applications. *Journal of Materials Chemistry*, **19**, 150–158.
[3] van Wazer, J.R. (1958) *Phosphorus and its Compounds*. New York: Interscience.
[4] Brow, R.K. (2000) Review: the structure of simple phosphate glasses. *Journal of Non-Crystalline Solids*, **263**, 1–28.
[5] Vogel, W. (1994) *Glass Chemistry*. 2nd edn. Heidelberg: Springer.

[6] Walter, G., Vogel, J., Hoppe, U. and Hartmann, P. (2003) Structural study of magnesium polyphosphate glasses. *Journal of Non-Crystalline Solids*, **320**, 210–222.

[7] Kasuga, T., Hosoi, Y., Nogami, M. and Niinomi, M. (2001) Apatite formation on calcium phosphate invert glasses in simulated body fluid. *Journal of the American Ceramic Society*, **84**, 450–452.

[8] Bunker, B.C., Arnold, G.W. and Wilder, J.A. (1984) Phosphate-glass dissolution in aqueous solutions. *Journal of Non-Crystalline Solids*, **64**, 291–316.

[9] Abou Neel, E.A., Pickup, D.M., Valappil, S.P. *et al.* (2009) Bioactive functional materials: a perspective on phosphate-based glasses. *Journal of Materials Chemistry*, **19**, 690–701.

[10] Brauer, D. S., Karpukhina, N., Law, R. V., and Hill, R. G. (2010) Effect of TiO_2 addition on structure, solubility and crystallisation of phosphate invert glasses for biomedical applications. *Journal of Non-Crystalline Solids*, **356**, 2626–2633.

[11] Lakhkar, N.J., Abou Neel, E.A., Salih, V. and Knowles, J.C. (2009) Strontium oxide doped quaternary glasses: effect on structure, degradation and cytocompatibility. *Journal of Materials Science: Materials in Medicine*, **20**, 1339–1346.

[12] Gentleman, E., Fredholm, Y.C., Jell, G. *et al.* (2010) The effects of strontium-substituted bioactive glasses on osteoblasts and osteoclasts in vitro. *Biomaterials*, **31**, 3949–3956.

[13] Salih, V., Patel, A. and Knowles, J.C. (2007) Zinc-containing phosphate-based glasses for tissue engineering. *Biomedical Materials*, **2**, 11–20.

[14] Ishikawa, K., Miyamoto, Y., Yuasa, T. *et al.* (2002) Fabrication of Zn containing apatite cement and its initial evaluation using human osteoblastic cells. *Biomaterials*, **23**, 423–428.

[15] Navarro, M., Ginebra, M.P., Clement, J. *et al.* (2003) Physicochemical degradation of titania-stabilized soluble phosphate glasses for medical applications. *Journal of the American Ceramic Society*, **86**, 1345–1352.

[16] Parsons, A.J., Ahmed, I., Haque, P. *et al.* (2009) Phosphate glass fibre composites for bone repair. *Journal of Bionic Engineering*, **6**, 318–323.

[17] Brauer, D.S., Russel, C., Vogt, S. *et al.* (2008) Degradable phosphate glass fiber reinforced polymer matrices: mechanical properties and cell response. *Journal of Materials Science: Materials in Medicine*, **19**, 1 21–127.

[18] Pessani, J.P., Al-Ibrahim, N.S., Rabelo Buzalap, M.A. and Toumba, K.J. (2008) Slow-release fluoride devices: a literature review. *Journal of Applied Oral Science*, **16**, 238–246.

[19] Featherstone, J.D.B. (2000) The science and practice of caries prevention. *Journal of the American Dental Association*, **131**, 887–899.

[20] Kendall, N.R., Jackson, D.W., Mackenzie, A.M. *et al.* (2001) The effect of a zinc, cobalt and selenium soluble glass bolus on the trace element status of extensively grazed sheep over winter. *Animal Science*, **73**, 163–169.

5

The Structure of Bioactive Glasses and Their Surfaces

Alastair N. Cormack

Kazuo Inamori School of Engineering, New York State College of Ceramics, Alfred University, Alfred, USA

5.1 STRUCTURE OF GLASSES

The atomic structure is everything to a glass, in that it affects all its properties, especially its bioactivity and degradation rate.

Glasses, including bioactive glasses, are non-crystalline, which means that their atomic scale structure is characterized by an absence of the translational periodicity associated with crystalline structures. This lack of long-range order means that traditional crystal structure concepts such as unit cells and lattice vectors cannot be used to discuss the structures of glasses. It also means that standard experimental probes of structure, such as X-ray or neutron diffraction, cannot be used to measure, or determine, the atomic arrangements in glasses. Fortunately, the majority of bioactive glasses contain a lot of silica, and so their structures can be described in the same way as other silicate glasses, for which quite a lot is known by analogy to crystalline silicates, whose structures have been well characterized in terms of the arrangement of

Bio-Glasses: An Introduction, First Edition. Edited by Julian R. Jones and Alexis G. Clare.
© 2012 John Wiley & Sons, Ltd. Published 2012 by John Wiley & Sons, Ltd.

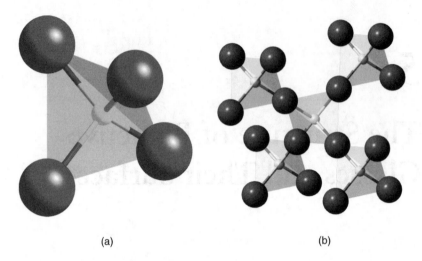

(a)   (b)

Figure 5.1 (a) A single SiO_4 tetrahedron and (b) four tetrahedra linked to a central tetrahedron through bridging oxygen ions.

the SiO_4 coordination tetrahedron. The tetrahedron is a geometric unit in which a silicon atom, sitting at its centre, is bonded to four oxygen atoms, which form the four vertices of the tetrahedron (Figure 5.1a); the tetrahedral arrangement comes about because of the strong, covalent character of the Si–O bond.

The starting point for the description of glasses usually begins, following Zachariasen [1], with vitreous silica, pure SiO_2, whose structure is considered to consist of a three-dimensional network of SiO_4 tetrahedra, each of which is linked to four other tetrahedra through an oxygen atom that is common to both tetrahedra. Each tetrahedron is associated with a node in the network. This is illustrated in Figure 5.1(b).

The network structure of silica is a random network, which is to say that there is no long-range order or translational periodicity, as indicated above. Such networks are usually characterized in terms of their ring size distribution. In a network, it is possible to define rings, whose size is determined by the number of nodes one traverses before arriving back at the node from which one started. Random networks will, in general, contain rings of all possible sizes, in contrast to crystalline structures, which, in general, contain rings of only one or two discrete sizes. For example, quartz, a crystalline polymorph of SiO_2, contains rings of six and eight tetrahedra, whereas cristobalite, a different silica polymorph, contains only six-membered rings (Figure 5.2).

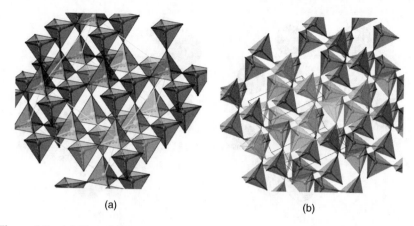

(a)   (b)

Figure 5.2   (a) Two different six-membered rings in the cristobalite structure and (b) a six-membered ring (pale gray/yellow) and an eight-membered ring (mid-gray/green) in the quartz structure. For a better understanding of the figure, please refer to the colour section (Figure 5).

When a glass contains components other than silica, as in the bioactive glasses, the network structure is modified; essentially, the network is disrupted (Figure 5.3). Some of the oxygen atoms are no longer bonded to two silicon atoms and are said to be non-bridging oxygens (NBOs). Essentially, these are created to charge-compensate for the addition of the modifier cations. In modified glasses, one now has to consider the distribution of NBOs. Is it possible for a tetrahedron to have only one NBO, or may some have more than one – perhaps as many as four? And is the corresponding distribution of the modifier cations homogeneous or heterogeneous? Since the presence of NBOs disrupts the links between nodes in the network (that is, some tetrahedra are no longer connected to each other), the ring size distribution will also change as modifiers are introduced into the structures. The distribution of the NBOs over the tetrahedral structural units is characterized in terms of a quantity denoted Q_n. Here n refers to the number of bridging oxygen atoms on the tetrahedron, so Q_3 represents a tetrahedron with one NBO, Q_2 units have two NBOs, and so on. Another way to characterize this is in terms of network connectivity (NC), which measures the average number of bridging oxygens on each tetrahedron (Chapter 2). Thus, at one extreme, silica has an NC value of four, and a structure comprising strings of tetrahedra would have an NC value of two. Because NC is an average quantity, it masks some actual variations in the distribution of

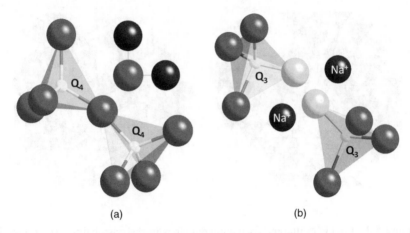

(a)   (b)

Figure 5.3 (a) Two tetrahedra (Q_4 species) with a common, bridging, oxygen ion are shown, along with an Na_2O molecule. (b) The two tetrahedra are now both Q_3 species, and each has a non-bridging oxygen ion, as the Si–O–Si bond is disrupted by the addition of Na_2O.

NBOs, but nevertheless NC has been found to be useful in predicting some properties, particularly the tendency of a glass toward bioactivity (see below and Chapter 2).

In addition to the properties of the network formed by the tetrahedra, other structural considerations include interatomic bond lengths and bond angles, and the coordination number of the ions, particularly the cations, but also the oxygen anions. In vitreous silica (without any network modifiers added), the Si–O bond length has a very narrow distribution because of the strength of the bond, and is around 1.6 Å. In addition, the O–O bond length does not vary much because of the tetrahedral shape of the SiO_4 unit, which causes the O–Si–O bond angle to be fixed at 109.4°, the internal tetrahedral angle. What does vary, however, is the angle through which adjacent tetrahedra are related, the Si–O–Si bond angle.

5.2 STRUCTURE OF BIOACTIVE GLASSES

Melt-derived bioactive glasses contain considerable amounts of modifiers: the composition of the eponymous Bioglass® is (in mol%) 24.35 Na_2O, 26.9 CaO, 2.57 P_2O_5, and 46.1 SiO_2. In addition, note the small, but critical, phosphorus content. The coordination number of phosphorus in the glass is four, and it is found in a tetrahedral environment, like silicon. This makes sense, as phosphate can also form a glass network

(Chapter 4). How the PO_4 tetrahedra are incorporated into the silica tetrahedral network is a matter of discussion, to which we will return later. Because of the amount of Na and Ca in these glasses, there is a large concentration of NBOs. Their number is too large to be distributed simply one to a tetrahedron, and so the structures contain Q_3, Q_2, Q_1 and Q_0 units. The PO_4 tetrahedra are particularly prone to having four NBOs, and, when they do, are called orthophosphate units. These structural units, along with their SiO_4 counterparts, are believed to play an important part in the bioactivity of the glasses.

The more NBOs that are present in a glass, the more depolymerized becomes its tetrahedral network. Bioactive glasses thus tend to be quite depolymerized, and it has been argued, for example, that a degree of depolymerization corresponding to an NC of around two is optimum for bioactivity (larger NC leading to reduced bioactivity). This is related to the formation of the hydroxyapatite layer on the glass surface, which requires the (partial) dissolution of the structure, and so is facilitated when the structure is not completely polymerized, because fewer Si–O–Si bonds need to be broken (Chapter 2). Recall that an NC of two corresponds, on average, to a structure containing just strings of tetrahedra in which each tetrahedron is connected only to two other tetrahedra; in other words, NC = 2 represents a structure that is adequately depolymerized, or fragmented, to promote hydroxyapatite formation.

5.3 COMPUTER MODELING (THEORETICAL SIMULATION) OF BIOACTIVE GLASSES

Although experimental probes of glass structure are limited in the information that they can provide, atomistic computer simulations have played a significant role in developing our current understanding of the structure of glasses, particularly silicates, including bioactive glasses. For a recent review, see [2].

Figure 5.4 shows the tetrahedral structure of a simulated bioactive glass, 45S5 Bioglass®, when all of the modifier cations (Na and Ca) have been removed from the picture. It can readily be seen that, in addition to the isolated orthophosphate groups, there are orthosilicate groups and P in Q_1 sites, as well as Q_3 and Q_4 Si sites. Naturally, the presence of the modifier cations, Na and Ca, as well as the NBOs, causes an increase in the variety of interatomic bond lengths that are needed to describe the structure. This makes it even more problematic to extract such

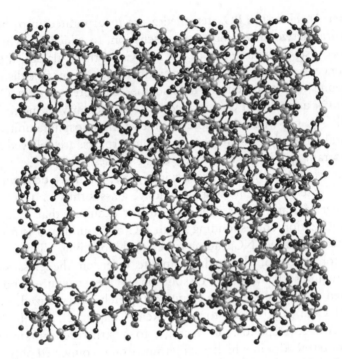

Figure 5.4 A snapshot of simulated 45S5 Bioglass, with the Na and Ca ions removed. Although many of the PO$_4$ groups are not connected to the tetrahedral network, there are some P–O–Si bonds. Si are pale gray/yellow, O are dark gray/red, and P are larger and mid-gray/purple. For a better understanding of the figure, please refer to the colour section (Figure 6).

information from scattering experiments. Recall that scattering data are 1D in nature, being a projection of the 3D structural information. It is very hard, if not impossible, to recreate the 3D structure from the 1D scattering information. (Note that, if one is dealing with crystalline materials, this is not the case, because the Rietveldt method of analyzing powder diffraction data is available to one.)

The simulations support the idea that bioactive glasses have a highly fragmented network structure, although the nature of the fragmentation at NC = 2 is more complex than simply chains of tetrahedra. Figure 5.5 shows the network fragments remaining after the largest fragment has been removed from the structure in Figure 5.4. The simulations also show that, as expected, in the most bioactive glasses there are a large number of orthophosphate groups – those not bonded to any other tetrahedra – and a significant number of orthosilicate groups as well. However, not all of the PO$_4$ groups are found to be Q$_0$ species: some are

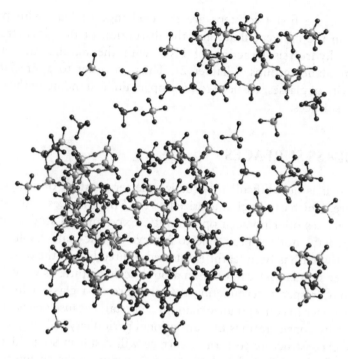

Figure 5.5 The tetrahedral fragments remaining when the largest fragment is removed from the structure shown in Figure 5.4. For a better understanding of the figure, please refer to the colour section (Figure 6).

connected to the SiO_4 tetrahedral network, which (because of the Q_0 and Q_1 species) is more connected than an NC value of two, that is, a simple collection of strings of tetrahedra, would imply. The fact that some PO_4 units are not Q_0 means that there must be some P–O–Si bonds, and this point is somewhat controversial. While solid-state nuclear magnetic resonance (NMR) experiments can provide relative proportions of Q units, the proportions are averaged over the sample, giving a binary distribution of Q_n species; the computer simulations indicate a more diverse distribution, and this is supported by other spectroscopic data. It should, perhaps, be noted that P–O–Si are relatively rare in other compounds, but are not unknown.

While we have a reasonable picture of the atomic arrangement in bulk bioactive glasses, and can correlate, loosely, the degree of fragmentation of the tetrahedral network with the degree of bioactivity, the actual structural features that initially control the bio-reactivity are those on the surface of the biomaterial. Recall Hench's sequence of reaction processes: the first two steps are the exchange of Na^+ with proton species in the ambient fluid, and the disruption of the glass structure through hydrolysis (reaction of H_2O with the tetrahedral network to form silanol groups, Chapter 2). Thus, in order to appreciate the bioactivity of glasses, we need to develop an understanding of the nature of their surfaces.

5.4 GLASS SURFACES

Whereas it is not difficult to visualize the termination of a crystalline structure, which can be characterized by the Miller indices of the surface plane, glasses offer no such opportunity. Experimental studies of glass surfaces have not yet been able to achieve atomic-scale resolution, so much of what has been learnt is inferred indirectly. On the other hand, computer simulations can offer an atomic-scale picture, as we will see.

In some respects, terminating the structure of a glass at its surface may be easier to conceptualize than for a crystal, because one is not constrained by considerations of translational periodicity. 'Slicing' through a bulk glass structure to form a surface will result in some 'dangling' bonds, which are Si–O bonds that were cut during the slicing process. Thus, the electrons in an Si–O bond suddenly find themselves without an oxygen with which to be shared. This is not an energetically favored state, so Nature will try to rearrange the structure to remove such disrupted bonds, and to restore, where possible, the optimum coordination environment of the silicon cations. The consequence is that the tetrahedral network structure at the surface will contain tetrahedra connected in smaller sized rings than in the bulk. In some cases, the tetrahedral coordination geometry cannot be restored – that is, sometimes it is not possible for a silicon ion to find four conveniently located oxygen atoms, and it has to be satisfied with just three. In such a case, the Si will not be in a tetrahedral geometry, but will form more of a planar coordination environment. In addition, in pure silica, NBOs may be formed at the surface, as depicted in Figure 5.6. All of these structural features – two- and three-membered rings, NBOs, and under-coordinated silicon – are defect structures, compared to the bulk structure, and are high-energy, reactive sites (recall that in bulk silica all of the oxygens are bridging). These are the surface features that will be the first to react with environmental species, such as water [3, 4].

In polycomponent glasses, the picture that we have of the surface structure is a bit more complicated, because, for example, they usually

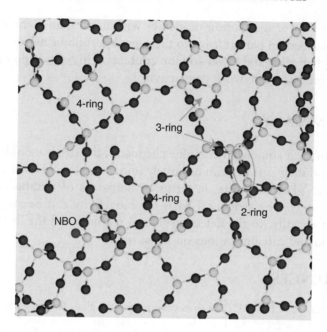

Figure 5.6 Structure of simulated surface of vitreous silica. Various structural features are identified, which are generally not seen in the bulk glass structure and are therefore considered to be defects.

already contain NBOs, so their appearance on the surface is no surprise. What is known is that there is an excess of sodium at the surface, with respect to the bulk, and this is the case, too, with bioactive glasses [5]. Also, the surface contains more smaller sized rings, particularly two- and three-membered rings, as well as under-coordinated silicon, as was found for pure silica, described in the previous paragraph. However, the fraction of two- and three-membered rings is lower in the more bio-reactive glasses than in the less bio-reactive glasses, because the larger amounts of modifiers present in the former allow the surface structure to relax more to bulk-like configurations than the higher silica-containing less bioactive glasses [6].

In addition, for bioactive glasses, the surface concentration of Na^+ seems to be greater than for bio-inert glasses, though this is primarily because of the higher bulk concentration of Na in the more bioactive glasses. There is an association between the Na^+ and the surface NBOs to form $Na^+ \cdots NBO$ pairs, sites that promote the dissociation of water through protonation of the NBOs. These sites, along with the under-coordinated silicon, which also promote protonation of the NBOs, provide for a more hydrophilic surface for the more bioactive glasses.

Thus, the slower dissolution of glasses with more than 60 mol% SiO_2 can be ascribed, in part at least, to their less hydrophilic nature, because of the higher silica and lower sodium contents. This is in addition to the effect of increasing network connectivity.

5.5 SUMMARY

In spite of their somewhat complex chemistry, the structure of bioactive glasses and their surfaces can be largely understood using basic concepts familiar to glass scientists, in terms of networks of tetrahedral coordination units. The structural basis for bioactivity can be ascribed, at least superficially, to the defect structural features and the distribution of sodium and calcium cations on the surface.

REFERENCES

[1] Zachariasen, W.H. (1932) The atomic arrangement in glass. *Journal of the American Chemical Society*, **54**, 3841–3851.
[2] Tilocca, A. (2009) Structural models of bioactive glasses from molecular dynamics simulations. *Proceedings of the Royal Society A: Mathematical Physical and Engineering Sciences*, **465**, 1003–1027.
[3] Du, J.C. and Cormack, A.N. (2005) Molecular dynamics simulation of the structure and hydroxylation of silica glass surfaces. *Journal of the American Ceramic Society*, **88**, 2978–2978.
[4] Tilocca, A. and Cormack, A.N. (2010) Surface signatures of bioactivity: MD simulations of 45S and 65S silicate glasses. *Langmuir*, **26**, 545–551.
[5] Tilocca, A. and Cormack, A.N. (2009) Modeling the water–bioglass interface by *ab initio* molecular dynamics simulations. *ACS Applied Materials and Interfaces*, **1**, 1324–1333.
[6] Tilocca, A. and Cormack, A.N. (2008) Exploring the surface of bioactive glasses: water adsorption and reactivity. *Journal of Physical Chemistry C*, **112**, 11936–11945.

6

Bioactive Borate Glasses

Steven B. Jung

MO-SCI Corporation, MO, USA

6.1 INTRODUCTION

Bioactive borate glasses are a subset of the bioactive glass family that includes silicate, borate and phosphate glasses. Traditionally, bioactive glasses are glassy materials that will react in an aqueous environment such as body fluids to form inorganic crystalline calcium compounds. Hydroxycarbonate apatite (HCA) is the compound that forms on most bioactive glasses and also happens to be very similar to the inorganic component of bone. Bioactive glasses will readily bond to connective tissue, such as bone and soft tissues, and they can also elicit biological responses, such as promoting cell migration and cell differentiation, and deliver ions necessary for the healing process (Chapter 2). An image of reacted bioactive borate glass converted to HCA is shown in Figure 6.1. Notice the porous nanocrystalline needle-like microstructure of the HCA.

This chapter will begin by discussing the typical bioactive borate glass reaction to form HCA and the reaction kinetics of borate glasses and how they differ from silicate glasses. Testing methods will be described in a general manner to discuss the pros and cons of *in vitro* and

---

Bio-Glasses: An Introduction, First Edition. Edited by Julian R. Jones and Alexis G. Clare.
© 2012 John Wiley & Sons, Ltd. Published 2012 by John Wiley & Sons, Ltd.

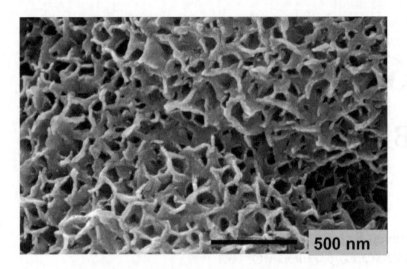

Figure 6.1   SEM image of HCA formed on bioactive borate glass reacted *in vivo*.

in vivo testing for reactive materials such as bioactive glass. Advances in bioactive borate glass technology will describe new methods of stimulating biological responses, such as bone and blood vessel formation (angiogenesis), without the use of expensive growth factors. A section will cover several applications of bioactive borate glasses in the medical field, and the last section will cover biomaterial design and provide insight into the design and development of new biomaterials and commercial products.

6.2 WHAT DIFFERENTIATES A BIOACTIVE BORATE GLASS FROM OTHER BIOACTIVE GLASSES?

Bioactive glasses that form HCA in general require calcium as a component (typically >10 wt%) and, beyond that, the composition is fairly open to customization. In the case of the bioactive borate glasses, the major glass former is B_2O_3 as opposed to SiO_2 or P_2O_5, and the compositions can contain an array of alkali metal (Li, Na, K, etc.), alkaline-earth (Mg, Ca, Sr, Ba, etc.) and transition-metal (Fe, Cu, Zn, Ag, Au) elements. A list of the silicate and borate glass compositions described in this and other chapters is shown in Table 6.1.

Bioactive glasses have been well documented to form a calcium phosphate layer when immersed in a phosphate-containing solution like simulated body fluid (SBF). The reaction that occurs in silicate glasses is described in detail in Chapter 2. Once the glass surface has been

Table 6.1 Compositions of some common bioactive glass (wt%).

Glass	B_2O_3	Na_2O	CaO	K_2O	MgO	SiO_2	P_2O_5	CuO	SrO	ZnO	Fe_2O_3
45S5	0	24.50	24.50	0	0	45.00	6.00	0	0	0	0
13-93	0	6.00	20.00	12.00	5.00	53.00	4.00	0	0	0	0
13-93B3	53.00	6.00	20.00	12.00	5.00	0	4.00	0	0	0	0
1B	17.00	24.00	23.90	0	0	29.30	0	0	0	0	0
2B	33.10	23.40	23.40	0	0	14.40	0	0	0	0	0
3B	48.60	22.90	22.90	0	0	0	0	0	0	0	0
Cu-3	52.79	5.98	19.92	11.95	4.98	0	3.98	0.40	0	0	0
CS	51.73	5.86	19.52	11.71	4.88	0	3.90	0.40	2.00	0	0
CSZ	51.20	5.80	19.32	11.59	4.83	0	3.86	0.40	2.00	1.00	0
CSZF	50.88	5.76	19.20	11.52	4.80	0	3.84	0.40	2.00	1.00	0.40

hydrated, the glass begins to dissolve, and a silica-rich layer forms on the surface of the glass. The low solubility of silica in body fluids is the reason for the silica layer formation. Calcium ions from the glass diffuse through the silica-rich layer and chemically bond with phosphate present in the body fluids, forming an amorphous calcium phosphate layer [1]. Over time, the calcium phosphate layer crystallizes into HCA. The formation of the HCA is what allows tissues like bone to chemically bond to the bioactive glasses [2].

Bioactive borate glasses, unlike silicate glasses, form HCA directly on the surface of the underlying unreacted glass [3], without forming a borate-rich layer. This is because the borate is readily soluble in body fluids, similar to the phosphate glasses (Chapter 4). The degradation products of the glass can be passed through the body naturally, predominantly through urine [4]. The lack of a diffusion layer allows borate glasses to react to completion without a significant reduction in the dissolution kinetics [5]. To establish how borate and silicate glasses differ in behavior, a set of glasses were designed based on the original 45S5 Bioglass® composition, where different proportions of the silica were replaced by borate. Figure 6.2(a) illustrates the weight loss (after reaction in dilute phosphate solution at 37 °C) from: 45S5 silicate (0B); two borosilicate glasses (1B and 2B), in which one-third and two-thirds of the silica in the 45S5 composition was replaced by borate, respectively; and a borate glass (3B), in which all the silica was replaced by borate. Only the borate glass (3B) dissolved fully to the theoretical weight loss (~58%), as the silica-containing glasses essentially stopped well short (35% and 42%, respectively).

The kinetics of each glass are modeled in Figure 6.2(b) and show that the 3B glass reaction is the most rapid to completion, and has reaction

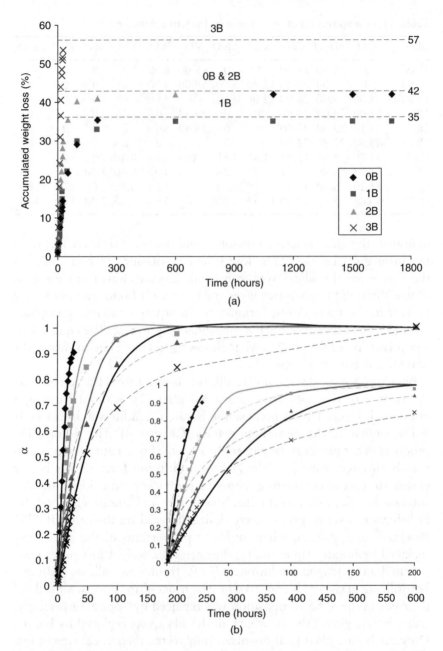

Figure 6.2 (a) Accumulated weight loss versus time for a silicate (0B), two borosilicates (1B and 2B), and a borate glass (3B) in 0.02 M K_2HPO_4 solution at 37 °C. (b) Kinetic analysis for the glasses shown in (a), where the solid line denotes the contracting volume model and the dashed line denotes the diffusion model. (Data acquired from Ref. [5].)

kinetics that can be modeled by the contracting volume model (surface dissolution). Each of the three glasses containing silica initially starts off with similar kinetics (solid line), but, owing to the formation of a silica-rich layer, the reaction of each glass slows and the data are fit by the slower diffusion model (dashed line). This means that the silica-rich layer is acting as the conduit for the dissolved ions from the glass to diffuse and either be released to the body fluids or form HCA. The slower diffusion kinetics for silicate glasses can be a distinct disadvantage in achieving full conversion of the bioactive glass, especially in larger glass particles (>300 μm) or monolithic implants. It has been documented that silicate-based bioactive glass implanted in human tooth sockets still had unreacted glass particles present four years after implantation [6]. Of course, the choice of the best glass will depend on the intended application. The rate of degradation or conversion to apatite that is required may differ depending on the clinical indication.

6.3 EVALUATING REACTIVE MATERIALS (*IN VITRO* VERSUS *IN VIVO* TESTING)

It is extremely challenging to create an *in vitro* test for potential medical devices that mimics the *in vivo* environment. One of the main difficulties is that body fluid flows *in vivo* and most *in vitro* systems are static. However, there are a host of other factors that are difficult to emulate, such as the mechanical forces that implants will experience, and the number of different cell types, proteins, and growth factors that are present in the blood.

Reactive materials such as bioactive glasses are especially difficult to evaluate *in vitro*, because the environment, such as ion concentration of the media and pH, can change drastically from the initial environment as the materials degrade, which can lead to false interpretation [7]. Much work has been published on the bioactive silicate glasses *in vitro* [8], but, as shown in Figure 6.2, these silicate glasses react one to two orders of magnitude slower than some of the borate glasses that will be described later in the chapter. This high rate of ion release makes cell culture studies with borate glasses difficult to evaluate. Traditional methods of cellular compatibility testing in static cell cultures will show in general that borate glasses are toxic, but this turns out to be a false negative [7].

In vivo evaluation of borate glasses shows the exact opposite of the *in vitro* tests, in that bone and soft tissue not only can survive in the presence of borate glass, but also can regenerate tissue just as effectively as silicate-based bioactive glasses [9]. The major difference between the *in vivo* and *in vitro* models is that the *in vivo* study is a fully dynamic system, with body fluids constantly being replenished and the pH of the micro-environment being much closer to the ambient. The culture medium in the static *in vitro* model will show a significant change in pH, usually increasing to about 9 or 10 for a borate glass, which would be toxic to cells. The ionic concentration of the medium can also be significantly altered depending on the ratio of glass to media used [7]. Therefore, *in vitro* testing is much more effective in evaluating chemically inert materials, where the pH and the culture medium content do not change significantly. Bioreactor systems comprising fluid flow may improve *in vitro* tests in the near future.

The problem is illustrated in Figure 6.3, which shows the total DNA of a bone cell line present after four days of growth in cell culture. Static culture is compared to dynamic culture, although here dynamic culture was a rocking motion to stimulate movement of ions rather than fluid flow. An increase in total DNA is an indication of an increase in the number of active cells. The data show that even rocking the culture had a significant effect in changing the results. In static culture, the increasing

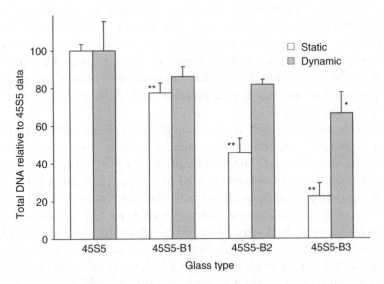

Figure 6.3 Total DNA of bone cells exposed to bioactive glasses in static and dynamic culture after four days for: silicate-based 45S5 Bioglass; two borosilicate glasses, 45S5-B1 (one-third of the silica replaced with borate) and 45S5-B2 (two-thirds of the silica replaced by borate); and a borate glass, 45S5-B3. The * and ** indicate statistically significant differences compared to 45S5 Bioglass. (Data acquired from Ref. [7].)

level of boron in the glass (increased reaction rate) caused cell death due to significant changes in the micro-environment just above the surface of the glass where the cells were attached. Using the same conditions except that the culture is gently disturbed by periodic agitation, the amount of DNA present in the culture was higher, and only the full borate glass is statistically lower than the silicate glass [7]. This improvement of cell sustainability in the dynamic culture over the static culture indicates that, in a fully dynamic system, such as the body, the expected toxicity of the borate glass could potentially be non-existent.

The main point from this section is to understand that different types of materials require different types of testing to get an accurate assessment of how the material will react when implanted in a mammal. Having an understanding for the material and how it reacts in an aqueous environment is valuable information when planning an experiment. It is important to know that sometimes materials can fail tests, and sometimes tests can fail materials.

6.4 MULTIFUNCTIONAL BIOACTIVE BORATE GLASSES

Traditionally, bioactive glasses have been used in orthopedic applications for the regeneration of bone [10, 11]. Bone is a highly vascular material, and, in fact, bone cells cannot survive unless they are within $100-200\,\mu m$ of a blood capillary [12]. Emphasis for most bioactive glass testing is typically on the ability to form HCA and have osteoblast-like cells differentiate into osteocytes [8]. These two parameters are important, but the formation of soft tissues like blood vessels is just as important in promoting bone growth *in vivo*, especially in full-thickness segmental defects, where the surrounding vascular network has been severely damaged [12].

Administration of growth factors such as vascular endothelial growth factor (VEGF) has been widely studied as a method for inducing vascular growth in bone implants [13]. Unfortunately, growth factors typically are expensive (thousands of dollars per dose) and are not likely to be the long-term solution for this problem. Fortunately, the body uses ions such as copper to regulate angiogenesis, and copper can easily be added to bioactive glasses for slow and controlled release to the surrounding tissues [9, 14]. An added advantage of using bioactive glass is that the release can occur over the course of weeks to months, which is preferable to the relatively fast release of a growth factor simply coated on an implant material [9].

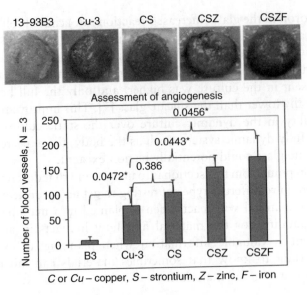

C or Cu – copper, S – strontium, Z – zinc, F – iron

Figure 6.4 Assessment of angiogenesis by scaffolds of borate glasses doped with metal ions that were implanted in rats for six weeks. Copper and zinc were the elements that promoted statistically significant increases in vascularity. (Data acquired from Ref. [8].)

Angiogenic properties of bioactive glass are also important to soft tissue regeneration. In India, it is estimated that there are more than 50 million people who suffer from diabetes. The USA, Europe, and most of the developed nations also have millions of people with diabetes. Most of these people at some time in their lives will develop wounds that are caused by diabetes or an associated disease that has delayed healing or will not heal at all. These wounds are typically vascular-deficient and will only heal when the vascular network has been repaired. Angiogenic bioactive borate glasses, like those shown in Figure 6.4, are currently being investigated for healing these non-healing diabetic wounds in humans, and the results are quite promising [15].

Small additions (a small weight percentage) of certain metal ions like copper and zinc can have statistically significant increases on the number of blood vessels adjacent to bioactive glasses. The scaffolds shown at the top of Figure 6.4 change from almost white (13-93B3) with relatively few blood vessels to dark purple (CSZF), and this acts as a visual indication to the increase in blood vessels present in the scaffolds. More will be covered on this application at the end of the chapter.

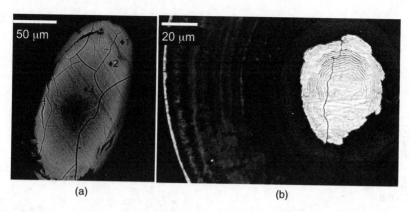

Figure 6.5 (a) Cross-section of a 13-93B3 bioactive borate glass fiber converted to HCA after six weeks *in vivo*. (b) Cross-section of a CSZF fiber converted *in vivo* for six weeks that formed HA at the outer edge and calcite (white) at the centre. (Images modified from Ref. [8].)

A new phenomenon in bioactive glass conversion has been discovered with the bioactive borate glass family. As mentioned earlier, calcium-containing bioactive glasses react to form HCA when implanted in a phosphate-containing solution such as body fluids. It has been discovered that the addition of certain elements can shift the reaction from HCA to calcium carbonate [9]. This shift may seem undesirable, but in fact calcium carbonate is a precursor to HCA, and can be dissolved by osteoclasts at a higher rate than HCA [16]. Theoretically, this means that the body can remodel the biomaterial into natural tissue at a higher rate, leaving behind no evidence that an implant was ever there.

Figure 6.5 shows scanning electron microscope (SEM) cross-sections of two bioactive borate glass fibers that have been implanted *in vivo* for six weeks. The undoped 13-93B3 glass (a) formed HCA from the outer edge to the centre. The CSZF glass (b), which has been doped with copper, strontium, zinc, and iron, formed a thin (2–3 μm) layer of HCA at the outer edge, but then the remaining material (bright white) was identified as calcium carbonate in the form of calcite by X-ray diffraction (XRD) [9].

This phenomenon may be partially explained by the schematic shown in Figure 6.6. The size of the dopant and the dopant concentration have been identified *in vitro* [17] and *in vivo* [9] to have caused inhibition of HCA formation. Atoms with ionic radii smaller than ~0.8 Å appear to have moderate to strong inhibition on HCA formation depending

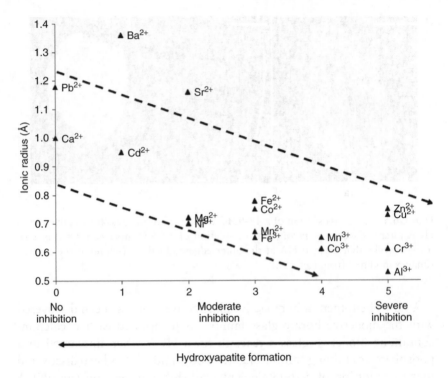

Figure 6.6 Schematic showing the inhibitory effects of metal ions on HCA formation on borate glasses.

on concentration. One theory behind the HCA inhibition is that the addition of small ions to the HCA structure acts as road blocks for the formation of the next unit cell, essentially stopping HCA or other calcium phosphate formation [9]. The formation of calcite can be explained by calcium salt solubility. Calcite has a solubility limit just above HCA [18], but, if calcium phosphate phases cannot form, then any free calcium in a carbonated solution like body fluids will crystallize to the calcite phase of calcium carbonate.

6.5 APPLICATIONS OF BIOACTIVE BORATE GLASSES IN ORTHOPEDICS AND DENTAL REGENERATION

Silicate-based bioactive glasses have been used in treating bone defects for over 20 years [8]. There are several articles in the literature describing the need for silica to stimulate gene expression with bone grafting materials [19]. Some of the work that is shown here will challenge that theory,

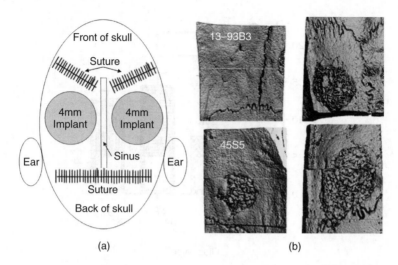

(a)   (b)

Figure 6.7 (a) Schematic of the rat skull showing the placement of the implants. (b) Comparative X-ray microtomography images of 13-93B3 fiber scaffolds and 45S5 Bioglass particles after 12 weeks *in vivo*. The side of the CT image that is labeled is the bottom side of the skull, and the unlabeled is the top of the skull. The bottom side of the 13-93B3 scaffold is completely covered over with new bone and cannot be seen. (Images acquired from Ref. [8].)

because silica-free bioactive borate glasses have been shown to stimulate bone just as well as the silicate-based 45S5 Bioglass in rat calvaria defects (Bi, L., Jung, S.B., Day, D.E., *et al.*, unpublished) [9]. Borate glass (13-93B3) and 45S5 were compared in an identical bone growth model to determine the effect that each has on bone regeneration. The model used was the critical sized rat calvarial defect (4 mm), a schematic of which is shown in Figure 6.7(a). After 12 weeks, X-ray microtomography was used to visualize the defects, and a representative image from each glass is shown in Figure 6.7(b). Although it is difficult to compare particles with fibers, it was observed that the borate glass was completely covered with new bone on the bottom side of the scaffold, whereas 45S5 Bioglass was not.

Histological measurements of bone growth across the centre of the implants ($N = 4$) are shown in Figure 6.8. The 45S5 and 13-93B3 are statistically similar, but the borate glass is completely devoid of silica. The general consensus among those who study bioactive glasses is that the presence of silica in bioactive glass is one of the most important components for stimulating bone growth. The present data are contradictory to the general consensus, in that a silica-free glass was statistically just as beneficial *in vivo* as the well-known 45S5 glass.

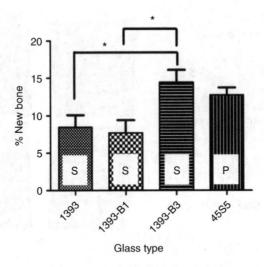

Glass type

Figure 6.8 Histological assessment of bioactive glass scaffolds implanted in rat calvaria for 12 weeks: 45S5 was implanted as particles; 45S5 and 13-93 were statistically similar, while the 13-93B3 had a statistically significant increase in bone growth over silicate 13-93 and a borosilicate glass 13-93B1 ($N = 4$). (Data acquired from Ref. [8].)

There is no question that soluble silica has been proven to up-regulate bone cells to promote osteogenesis, and the point of inclusion of the analysis is not to argue that point. The point is to show that perhaps there is something else going on with regards to the relationship between glass compositions and biological response.

6.6 SOFT TISSUE WOUND HEALING

Soft tissue wounds are an increasing problem worldwide as the population, especially in developed countries, continues to age. The problems associated with diabetics with slow-healing wounds is leading to expensive prolonged care by outpatient facilities, and non-healing wounds are leading to amputations. Not only are the wounds devastating to patients, but also the operations and prolonged care are expensive. These costs are a growing problem at the same time as governments are cutting healthcare benefits. Better treatment methods are needed to help to heal the wounds of these millions of people who have no other options but to live with the wounds or to have entire limbs removed.

Diabetic wounds often have vascular-deficient tissue, and when these tissues are damaged, the body has no effective way of supplying the natural growth factors and nourishment required to heal. This is why

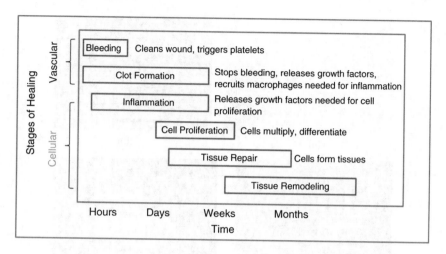

Figure 6.9 The stages of healing for connective tissue as a function of time.

diabetics can develop chronic non-healing wounds that may show no improvement for years. The healing process in a healthy person has two main stages of healing, first vascular and then cellular, as described in Figure 6.9. The initial stage (vascular) sets the wound up for the entire healing process, and without bleeding and the formation of a blood clot, the wound will not heal effectively.

The blood clot is the building block that stops bleeding, then the platelets release growth factors, while neutrophils and macrophages are recruited to clean the wound and release growth factors [20]. Eventually, epithelial cells migrate from the edges of the wound and form new skin. In a healthy person, this occurs in a well-defined process in which the wound closes and eventually disappears. A diabetic person may have health conditions that would disrupt this healing process, and the wound is essentially stuck and will not progress.

Bioactive borate glass nanofibers in the form of a pad have proven to be effective in healing chronic wounds. Figure 6.10 shows an image of the borate glass nanofiber and how it is applied to a wound. The fiber is flexible, similar to cotton to the touch, and can easily be applied to an open wound. The fiber is kept in place for three to four days and changed with periodic dressing changes by simply washing away any remaining fiber with sterile saline. A new dressing is applied and covered with an external bandage.

The first example of a chronic wound treated with the borate glass nanofiber is shown in Figure 6.11. The wound was on the patient's heel and was caused by constant pressure from lying in bed (pressure ulcer).

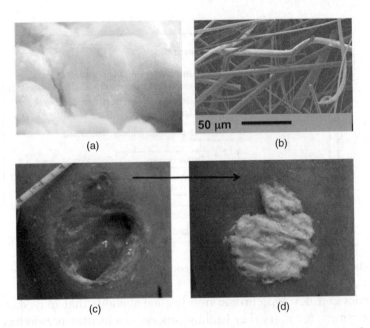

Figure 6.10 (a) A wad of bioactive borate glass nanofiber, (b) a high-magnification SEM image of the fiber, (c) a wound prior to dressing with the borate glass nanofiber, and (d) the same wound after dressing with the borate glass nanofiber. For a better understanding of the figure, please refer to the colour section (Figure 7).

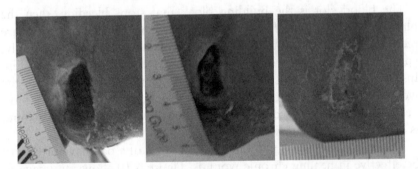

Figure 6.11 A chronic heel wound treated with borate glass nanofiber. Prior to treatment, the wound had existed for two years.

The patient had the wound for over two years, was a diabetic, and in relatively poor health. After approximately one month of treatments with the borate glass fibers administered twice per week, the wound had resolved.

A second example is a venous stasis ulcer on the upper leg of an obese woman in relatively poor health (prediabetic). The wound (Figure 6.12) was initially presented with a strong odor (likely a yeast infection) and

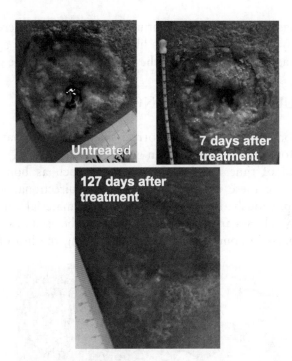

Figure 6.12 Large wound located on the upper leg of an obese woman. The wound was treated with bioactive borate glass nanofibers and was resolved in about four months. For a better understanding of the figure, please refer to the colour section (Figure 8).

was large in both diameter (~7 cm) and depth (2.5 cm). After just two treatments with borate glass fibers, the smell was significantly diminished (the glass is antimicrobial) and the wound had decreased in volume by ~50%. After four months the wound was fully resolved, with little scarring noted.

Lack of scarring is important, for two reasons: people do not want to look disfigured after their wound has healed; and scar tissue is only about 40% as strong as unscarred tissue, so wounds that heal but scar are likely to reopen with minimal abrasion.

Hemostasis (blood clotting) and antimicrobial properties go hand in hand when it comes to the design of bandages for critical wounds. Bioactive borate glasses in a nanofiber form can be applied to a bleeding wound, and the microstructure of the fiber mat acts like a cellular sieve to separate the solid content of the blood and the liquid fraction. This separation is highly effective in forming blood clots and is currently being researched for applications on the battlefield. The added benefit of bioactive glasses is they also are antimicrobial. This antimicrobial effect is not totally understood, but it is likely to be due to the spike in pH when the glass is hydrated due to the ions released from the glass.

6.7 TISSUE/VESSEL GUIDANCE

Bioactive borate glasses with the correct calcium content have a unique ability to form hollow cavities in particles or fibers in a relatively short period of time (weeks). Living tissue such as bone or blood vessels can use these channels as guides for directional growth. As mentioned previously, bone is a highly vascular material, and we know that bioactive glasses are biocompatible and can be used to stimulate the healing process in connective tissues. Harnessing the biocompatibility

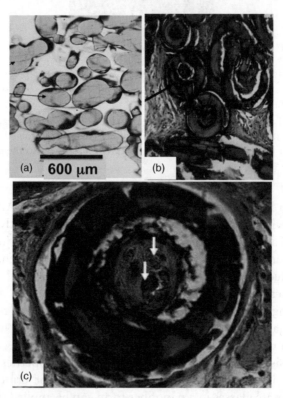

Figure 6.13 Bioactive borate glass forming hollow channels *in vivo*: (a) unreacted glass fibers, showing that they are originally dense glass fibers; (b) several reacted fibers that have tissue growing inside and around them; and (c) a single hollow fiber with blood vessels inside (arrows). For a better understanding of the figure, please refer to the colour section (Figure 9).

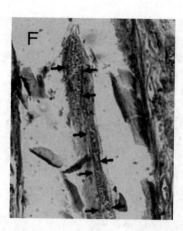

Figure 6.14 Borate glass fiber (F) that was cut parallel to the longitudinal axis. A blood vessel was present inside (arrows) growing the length of the fiber (total was ~500 μm).

with the final hollow form may be a new method of improving healing of full-thickness segmental bone defects or by acting as a vascular bridge from an area of high vascular concentration to low.

Figure 6.13 shows an example of solid glass fibers that were implanted *in vivo* for four weeks and formed hollow fibers. Soft tissue was present inside one of the hollow fibers and blood vessels were identified as well. Figure 6.14 shows a fiber that was cut parallel to the longitudinal axis and a blood vessel was found growing inside as further evidence to the examples in Figure 6.13. The blood vessel (arrows) was ~500 μm in length, and had visible red blood cells inside the vessel. In color, the red blood cells were stained fluorescent green.

6.8 DRUG DELIVERY

In situ drug delivery is an important concept for improving the healing process while only treating the site of interest. Drugs taken orally or intravenously are treating the entire body. Releasing pain killers, growth factors or medications at the needed site allows for potentially higher doses without harming the entire body. In the case of pain medication, this reduces the chances of the patient becoming dependent on the drug. Bioactive borate glasses can be pre-reacted in a phosphate solution and be made into the same hollow materials described in the previous section. The HCA formed by the borate glass is porous and can be loaded with a drug, and, once implanted, the drug will permeate into the surrounding

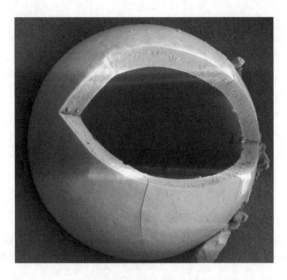

Figure 6.15 SEM image of a hollow HCA shell made by reacting a borate glass in a phosphate solution. The shell was intentionally broken to show the hollow centre. The shell diameter is ~150 µm.

tissues for days to weeks. An SEM image of the hollow HCA sphere is shown in Figure 6.15.

6.9 COMMERCIAL PRODUCT DESIGN

As a manufacturer of biomedical materials, and having developed many of the bioactive borate glass compositions discussed in this chapter, I can say that biomaterial design and product development are two totally different activities. They sound similar, but they are in fact very different. Biomaterial design is the development of a new material composition, perhaps a bioactive borate glass. This material can be characterized for several properties related to the material itself, or perhaps how it interacts biologically. Ideally, once the material composition is determined, it would be made into a product to serve some medical need. Depending on the application, the product could be a rigid scaffold or a flexible fiber mat.

The material composition should obviously be an effective component in whatever application is picked, but the form in which the material is delivered is what ultimately will sell the device. This may seem odd, since in university we are taught about material properties and how to maximize these for our application. So the science is what should sell the product – right? In reality, the science is only part, and likely a small part, of why practitioners choose a specific medical device. Some biomaterial implants are made from a single material, but many are a component of a composite such as bone putty or some of the other composites discussed in this book. In many cases, the "science" is covered up with a pliable resin or a polymer for forming the material to the desired shape during surgery. Ultimately, a surgeon is going to decide whether or not they use your device based on cost, the form of the device, ease of use, and the expected outcome.

A specific biological response (i.e. bone bonding) is the reason that a specific biomaterial or bioactive glass is chosen initially. However, ease of use is likely what a doctor cares about most. No one wants a great material that is impossible to use. A moderately effective material with excellent handling characteristics will sell much better and be used more often in surgery. Understanding the clinical need and addressing how doctors would ideally apply the material most effectively is the first thing to find out when developing a new biomaterial product. The second step is to get the material of interest into that form. When the product is what the clinician desires, the product has a greater chance of success.

Something about biomaterial design and product development that is often ignored in university is cost. For instance, say we were developing an electrical wire. Assuming that conductivity was most important, a material like gold or silver might be the best choice, because of their high electrical conductivity. Realistically, we also know that we will need relatively low cost. Therefore, a much cheaper material with lower, but still acceptable, conductivity – like copper – becomes the material of choice. The same is true for making biomaterials.

Cost-effective materials and methods are imperative to scale the product to a level that is sufficient for demand. What can be done on a bench top is not necessarily how something will be scaled for production. For example, we may have a two million dollar machine that can produce 8 g of viable material a day. This is not a scalable process, because the cost of the material would not be competitive, regardless of the biological outcome, and the only way to increase output is by buying more machines. In the described model, equipment overhead costs would be the major hurdle to overcome. Ideally, the idea could be closely mimicked by a much less expensive method of production, yielding a similar product with similar biological outcomes at a mere fraction of the cost.

Ease of manufacture is another issue often not covered in typical biomaterial courses. Manufacturing yield is a hidden cost in material processing that can have a significant impact on product success. For instance, an effective device is identified as a candidate for a new product, but the material is relatively expensive, and after processing yields only 10% of the starting material into saleable product. The question that should now be asked is this: Does adding a multiple of 10 to the cost of the product provide commercial success? If the clinical outcome is only marginally better than that achieved using an existing product that costs half as much as the proposed product, the likelihood of success is low.

Hopefully, this section has shed light on real-world issues with the development of biomaterials beyond the bench top and has helped you in evaluating methods of processing and where some hidden costs come into play. If you are really interested in knowing what an orthopedic surgeon thinks about the current bone augmentation products, and what they might like to see in a new product, it is best to find such a surgeon and ask. You can save a lot of development time and money by solving real problems and avoiding trying to solve problems that you yourself perceive to exist.

6.10 SUMMARY

This chapter has demonstrated several applications and was intended to provide an introduction to bioactive borate glass technology. This area of bioactive glass science is still in its infancy, so much of the science is still being uncovered. With many exciting results, such as bone growth and the successful healing of diabetic wounds, the future of bioactive borate glasses looks bright. With that in mind, not every material is ideal for every application, so understanding your material requirements and the costs associated with scaling the research to support commercial products is essential in picking a material for a proposed application.

REFERENCES

[1] Hench, L.L. (2006) The story of Bioglass. *Journal of Materials Science: Materials in Medicine*, 17, 967–978.
[2] Hench, L.L. and Paschall, H.A. (1973) Direct chemical bond of bioactive glass-ceramic materials to bone and muscle. *Journal of Biomedical Materials Research*, 4, 25–42.
[3] Marion, N.W., Liang, W., Reilly, G. *et al.* (2005) Borate glass supports the in vitro osteogenic differentiation of human mesenchymal stem cells. *Mechanics of Advanced Materials and Structures*, 12, 1–8.
[4] Gorustovich, A.A., Steimetz, T., Nielsen, F.H. and Guglielmotti, M.B. (2008) A histomorphometric study of alveolar bone modelling and remodelling in mice fed a boron-deficient diet. *Archives of Oral Biology*, 53, 677–682.

[5] Jung, S.B. and Day, D.E. (2009) Conversion kinetics of silicate, borosilicate, and borate bioactive glasses to hydroxyapatite. *Physics and Chemistry of Glasses*, **50**, 85–88.

[6] Suzuki, K., Misch, C., Arana, G. *et al.* (2011) Long-term histopathologic evaluation of bioactive glass and human-derived graft materials in Macaca Fascicularis mandibular ridge reconstruction. *Implant Dentistry*, **20**, 318–322.

[7] Brown, R.F., Rahaman, M.N., Dwilewicz, A.B. *et al.* (2008) Effect of borate glass composition on its conversion to hydroxyapatite and on the proliferation of MC3T3-E1 cells. *Journal of Biomedical Materials Research*, **88A**, 392–400.

[8] Hench, L.L., Hench, J.W. and Greenspan, D.C. (2004) Bioglass: a short history and bibliography. *Journal of Australian Ceramic Society*, **40**, 1–42.

[9] Jung, S.B. (2010) Borate based bioactive glass scaffolds for hard and soft tissue engineering. PhD thesis. Materials Science and Engineering, Missouri University of Science and Technology, Rolla, MO.

[10] Sanus, G.Z., Tanriverdi, T., Kafadar, A.M. *et al.* (2005) Use of Cortoss for reconstruction of anterior cranial base: a preliminary clinical experience. *European Journal of Plastic Surgery*, **27**, 371–377.

[11] Larsson, S. (2006) Cement augmentation in fracture treatment. *Scandinavian Journal of Surgery*, **95**, 111–118.

[12] Rouwkema, J., Rivron, N.C. and van Blitterswijk, C.A. (2008) Vascularization in tissue engineering. *Trends in Biotechnology*, **26**, 434–441.

[13] Leach, J.K., Kaigler, D., Wang, Z. *et al.* (2006) Coating of VEGF-releasing scaffolds with bioactive glass for angiogenesis and bone regeneration. *Biomaterials*, **27**, 3249–3255.

[14] Harris, E.D. (2004) A requirement for copper in angiogenesis. *Nutrition Reviews*, **62**, 60–64.

[15] Jung, S.B. (2011) Wound healing power of glass. *Nanotech Insights*, **2**, 2–4.

[16] Monchau, F., Lefevre, A., Descamps, M. *et al.* (2002) In vitro studies of human and rat osteoclast activity on hydroxyapatite, beta-tricalcium phosphate, calcium carbonate. *Biomolecular Engineering*, **19**, 143–152.

[17] Madsen, H.E.L. (2008) Influence of foreign metal ions on crystal growth and morphology of brushite ($CaHPO_4 \cdot 2H_2O$) and its transformation to octacalcium phosphate and apatite. *Journal of Crystal Growth*, **310**, 2602–2612.

[18] Lide, D.R. (ed.) (2009) *CRC Handbook of Chemistry and Physics*. Boca Raton, FL: Taylor and Francis/CRC.

[19] Notingher, I., Verrier, S., Romanska, H. *et al.* (2002) In situ characterisation of living cells by Raman spectroscopy. *Spectroscopy – an International Journal*, **16**, 43–51.

[20] Clark, R.A.F., Ghosh, K. and Tonnesen, M. (2007) Tissue engineering for cutaneous wounds. *Journal of Investigative Dermatology*, **127**, 1018–1029.

7

Glass-Ceramics

Wolfram Höland

Ivoclar Vivadent AG, Schaan, Principality of Liechtenstein

7.1 GLASS-CERAMICS AND THEIR USES

The name "glass-ceramic" already tells us something about this fascinating material, which was invented in 1959. It is a combination of the two everyday words – "glass" and "ceramic." As we all know from familiar objects like drinking glasses and windows, glass is usually transparent. Ceramics are used in scientific and technical fields to make, among other things, electronic equipment and machines. We also use them in day-to-day life in the form of cups, mugs, and vases to name but a few.

In the first chapter of this book, we introduced you to the material we know as "glass." We learned that glass is a supercooled, frozen liquid, which, unlike ceramic, does not contain any crystals, and which is characterized by what is known as a transition range. We also learned that glass is made up of different molecular structures and microstructures (structures visible under a microscope). We were introduced to the special phenomenon of glass-in-glass phase separation, which occurs in microstructures. The fact that glasses can separate into different liquids

Bio-Glasses: An Introduction, First Edition. Edited by Julian R. Jones and Alexis G. Clare.
© 2012 John Wiley & Sons, Ltd. Published 2012 by John Wiley & Sons, Ltd.

(similar to oil and water), which can be frozen in this state so that they are preserved at room temperature, is truly amazing and permits us to make useful inhomogeneous glasses. We called this phenomenon phase separation in glasses.

The term "ceramic" is the name of another type of material. In contrast to glasses, ceramics contain only a very small amount of a glass phase. In scientific language, this amount is expressed as 1% by volume (1 vol%). Ceramics are mainly composed of crystals. They may be made of one or several types of crystals: scientists refer to these as crystal phases. These crystals determine the main characteristics of the ceramic. For example, ZrO_2 (zirconia) crystals produce very strong and tough ceramic materials. The flexural strength (resistance to bending) of zirconia ceramics is more than 1000 MPa. This means that 10 000 times atmospheric pressure must be applied before the material bends and breaks (1 atm ≈ 0.1 MPa). The fracture toughness of the ceramic, which is symbolized by the abbreviation K_{IC} and describes the ability of a material containing a crack to resist breaking, is about 4.5 MPa·m$^{1/2}$. Apart from having to be strong and tough, ceramics can also be required to have other characteristics like magnetism or biocompatibility. Biocompatible ceramics have been specially developed for use in the human body, for example, in the form of bearing surfaces in artificial hips and knees and teeth.

Glass and ceramics make up a wide range of materials. The term "glass-ceramics" refers to a material that combines the characteristics of both types of materials to produce an even better material [1]. This improvement is made possible because the new material contains both glass components (glass phases) and crystals (crystal phases). Glass-ceramics are made by first designing and making a glass and then heating it to high temperature. Some of the glass will crystallize to become a glass-ceramic. This gives a very intimate relationship between the glass and crystal phases, yielding impressive mechanical properties. The special glass from which glass-ceramics are produced is called a base glass. The remarkable thing about glass-ceramics is not the fact that they are a successful mixture of glass and crystals, which can be fired at temperatures around 1000 °C to create the final product, but that the formation and growth of crystals can be controlled in a glass.

Simply put, glass-ceramics are materials that are composed of one or more glass phases and one or more crystal phases, with the crystals having been formed in a base glass.

This definition is very general and you will want to know exactly what makes this process possible and why it was only discovered in 1959.

The first attempts to produce crystals in glasses were made a long time ago. However, the final product always ended up having worse properties than the original material. The uncontrolled way in which crystals formed in the base glass caused forces (known as uncontrolled residual stresses) in the material that then made it break easily. In 1959, Stookey, a ceramics specialist, discovered a way of controlling the way in which crystals form. In the process, he learned that the controlled formation of crystals in a base glass goes hand in hand with the controlled formation of extremely small phases, which we call nuclei. According to his discovery, the controlled formation of crystal phases is connected to the controlled formation of nuclei (the start of the crystallization process called nucleation). In other words, nucleation and crystallization are so closely linked that one cannot take place without the other.

Because controlled nucleation produced crystallites (very small crystals) in specific areas of the base glass, glass-ceramics with very different and special characteristics could be produced. As a result of this finding, the characteristics that had previously been known only for glasses or ceramics could be combined in one material. Later on, we will take a closer at look at these special properties and examine glass-ceramic materials in more detail. In addition, we will find out about glass-ceramics that transmit light almost as well as glass and do not change size with changing temperature. This is important for esthetic dental restorations. We will look at moldable glass-ceramics, which allow us to create a wide variety of shapes. Moldable glass-ceramics are also used in dentistry, where biocompatible materials are also very important. We will also talk about very strong glass-ceramics that are either shaped with computer-assisted tools or molded. In order to understand the special characteristics and ways of using these materials, we will take a close look at the methods of controlling crystallization in the glasses used in the manufacture of glass-ceramics.

7.2 METHODS USED FOR THE CONTROLLED CRYSTALLIZATION OF GLASSES

As you now know, a specific crystallization method needs to be followed if the finished glass-ceramic is to contain glass components as well as crystals. One method of controlled crystallization involves controlling the nucleation and crystallization processes within the base glass (also called internal processes or bulk processes), while in the other method, the crystallization process is controlled on the surface of the glass.

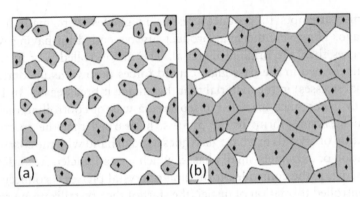

Figure 7.1 Controlled internal crystallization in glasses for producing glass-ceramics: (a) nucleation of crystallites in a glass; and (b) crystals grow and join at grain boundaries. (Reprinted with permission from [2]. Copyright (2006) W. Höland.)

In internal nucleation and crystallization, centers, or nucleation sites, from which the crystallization process starts are required. These sites are formed by chemical compounds that are added to the base glass. The chemical compounds, which are called nucleating agents, speed up the beginning of a process. We will explain what these nucleating agents are all about in the next few paragraphs. These substances form a crystalline phase, a foreign substrate, on which the actual crystal phase grows. This phenomenon is shown in Figure 7.1.

Glass-in-glass phase separation provides another way of starting the internal crystallization process within a base glass. During the separation of the glass into different phases, specific ions are concentrated into one phase, which enables nuclei and then crystal to form.

In addition to the method involving internal crystallization, the process of surface nucleation and crystallization is used to produce glass-ceramics. A diagram of the mechanism is shown in Figure 7.2. Glass is crushed to form a powder, and the surface of the glass powder grains is activated to produce crystals. These crystals then proceed to grow inwards from the surface of the glass particles.

For both methods to work, the chemical composition must be just right, and the base glass has to be heated to a very high temperature in the region of 1000 °C. By heating the glass, the processes of nucleation and crystallization are started and kept under control. The thermal energy (heat) makes molecules move around in the base glass and stimulates the growth of crystals in very specific areas. This is clearly shown in Figures 7.1 and 7.2. When the crystals have reached the

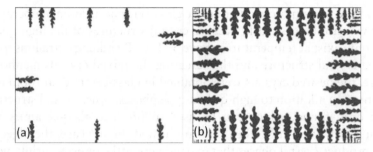

Figure 7.2 Controlled crystallization on the surface of the base glass in the development of glass-ceramics: (a) nucleation and (b) growth. (Reprinted with permission from [2]. Copyright (2006) W. Höland.)

desired size, the base glass is cooled down to stop them from growing further. The condition of the glass that has been achieved and the size of the crystals is preserved by freezing the material. This is how the microstructure of glass-ceramics is produced.

In the next section, we will look at different glass-ceramics with very special characteristics. We will explain their chemistry, describe the components that are used to start the crystallization process, and examine the crystallized parts of the material, which are responsible for producing the special characteristics of the glass-ceramic.

7.3 A GLASS-CERAMIC THAT HARDLY EXPANDS WHEN HEATED

The combination of SiO_2–Al_2O_3–Li_2O is suitable for making glass, which can be used to produce glass-ceramics that hardly expand when they are heated. This is important when making precision shapes, as the amount of expansion of other systems is difficult to control to exact measurements. These three oxides form what is called a glass formation system and can be supplemented with a wide range of other chemical compounds. The substances that cause the crystallization process (nucleating agents) are the most important compounds. TiO_2 and ZrO_2 are usually used for this purpose. When the base glass is heated, these compounds react and form a first crystal structure on which the desired crystal structure then grows when the glass is heated again. Over many years, researchers in the USA, Europe, and Japan developed crystallization processes in which SiO_2 mixed crystals (also called solid solutions) could be produced. The main aim of their work was to produce an SiO_2 crystal structure that would not expand when the material was heated and would remain present within the glass-ceramic at room temperature. This was not an easy task, because crystal structures of the high-quartz type only exist at temperatures above 573 °C. By adding extra ions to the mixed crystal structure and thus creating the mixed crystals mentioned above, the desired crystals were obtained in glass-ceramics at room temperature. In addition to high-quartz-type phases, other crystal structures have been found, which can be used to produce glass-ceramics that expand only very little when they are heated. In any case, the objective is to produce crystal phases that expand very little, or even shrink, when they are heated.

The glass-ceramic described above is characterized by the fact that it expands only minimally. Therefore, this type of material is used as a mounting for telescope mirrors, for microelectronic components (stepper motors for producing microchips are made of these materials), or for household goods (stovetops, baking dishes).

If the crystal structures are prevented from growing larger than 200 nm in size, the final glass-ceramic will be very translucent – in other words, it will allow a lot of light to pass through it. Consequently, a material can be produced that looks like glass, but expands only very little when it is heated.

7.4 HIGH-STRENGTH, MOLDABLE GLASS-CERAMICS FOR DENTAL RESTORATION

One of the main objectives in the development of new materials has always been to increase their strength. This is also valid for glass-ceramics. At first, efforts were made to produce very strong glass-ceramics by using the method of controlled surface crystallization. Even though a strong material was developed using this method, the ceramic objects made with this material were considerably weakened if their surface was damaged.

Therefore, alternative options had to be found and materials were developed by using the method of controlled crystallization within the glass. Even though they were not as strong as sintered technical ceramics (for example, ZrO_2), they were strong enough, and they could be molded into the desired shape. One of these glass-ceramics is based on a mixture of several substances consisting primarily of SiO_2–Li_2O. In this "multi-component system," metals or P_2O_5 are used as the substances that cause crystallization (nucleating agents). P_2O_5 in combination with a part of the Li_2O causes lithium phosphate phases to form, on which the desired main crystal structure made of lithium disilicate ($Li_2Si_2O_5$)

grows. When the plate-like (also called lath-like) crystals are heated, they grow into an interlocking microstructure. This process reinforces the material, so that its resistance to bending (flexural strength) is very high at 400–700 MPa. The large concentration of crystals (50 to 60 vol%) contained in the microstructure of the glass-ceramic is responsible for making the material much stronger than other glass-ceramics. However, as the material also contains some glass, it softens if it is heated to 920 °C. Therefore, it can be molded into different shapes. In a nutshell, the material combines the advantages of ceramics (high strength) with those of glass (moldability).

Both high strength and moldability are important features of the bio-compatible materials used to repair teeth. Very strong lithium disilicate glass-ceramics, for example, are popular for making dental crowns and bridges that do not need to be strengthened with metal. A very strong glass-ceramic is used to make a base. Another glass-ceramic material con-taining apatite crystals is applied on this base and shaped to resemble the original tooth. Figure 7.3 shows a bridge that has been made to replace

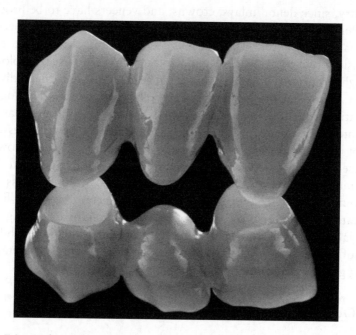

Figure 7.3 A dental bridge (on a mirror) made of biocompatible materials. A very strong glass-ceramic has been used to create the base. This framework has been coated with a glass-ceramic that looks like natural enamel. (Reprinted with permission from [2]. Copyright (2006) W. Höland.) For a better understanding of the figure, please refer to the colour section (Figure 10).

one missing tooth. It has a strong glass-ceramic base, which is coated with an apatite glass-ceramic. The amazing thing about these materials is not just their mechanical properties. Their optical properties are such that it is difficult to tell the difference between the glass-ceramics and natural teeth – they even have the right amount of chroma (products: IPS e.max® Press and IPS e.max® Ceram).

7.5 GLASS-CERAMICS THAT ARE MOLDABLE AND MACHINABLE

Glass-ceramics containing leucite crystals ($KAlSi_2O_6$) are produced by controlling the crystallization process at the surface of the glass. This method was discovered after that of internal nucleation. Even though this process was difficult to control at first, various methods have been developed that have resolved the problem. Recently, materials of this kind have become available that can be shaped by cutting or molding. This characteristic is beneficial in the biocompatible materials used in dentistry, since dental inlays, crowns, and veneers have to be produced in different shapes and sizes, depending on the tooth that needs to be repaired. In addition, these glass-ceramics must allow the dental professional who makes the tooth replacement to adjust the color and the light-transmitting ability of the materials to suit the individual patient. Of course, these materials have to be chemically and biologically compatible. The glass-ceramics mentioned combine all these character-istics and they can be produced in large quantities. Nevertheless, they cannot match the strength of lithium disilicate glass-ceramics. However, because they combine such a large number of favorable characteristics, these materials have a wide variety of uses (product: IPS Empress®).

Recent research results have shown that certain types of lithium disilicate glass-ceramics can also be shaped by molding or cutting. If the material is to be cut into the desired shape, it has to be done while it is in a soft condition. Afterwards, the shaped object is fired to obtain the strong glass-ceramic end product (IPS e.max® CAD).

7.6 OUTLOOK

Today, glass-ceramics with a large number of different characteristics and uses are available. Their variety is achieved by taking advan-tage of special mechanisms called controlled nucleation and controlled crystallization. Numerous new opportunities for using these materials are opening up every day.

REFERENCES

[1] Höland, W. and Beall, G.H. (2012) *Glass Ceramic Technology*, 2nd edn. Hoboken, NJ: John Wiley & Sons, Inc.

[2] Höland, W. (2006) *Glaskeramik*. vdf, Zürich, UTB, Stuttgart.

8

Bioactive Glass and Glass-Ceramic Coatings

Enrica Verné
*Department of Applied Science and Technology,
Politecnico di Torino, Torino, Italy*

8.1 INTRODUCTION

Bioactive glasses have been studied and used for a long time for a variety of medical applications, including small-bone substitutions, controlled drug delivery systems, bone cements, and generally for non-load-bearing implants [1]. However, their use has been limited by poor mechanical properties in tension or under cyclic loading. For any application for which good mechanical properties are needed, high-strength ceramics and metallic alloys are still the materials of choice. Alumina, zirconia, and some metallic alloys (Ti alloys, Co–Cr) are widely used in the biomedical fields owing to their very interesting mechanical properties. Specifically, alumina, zirconia, and their composites are used in several orthopedic and maxillofacial applications because of their biocompatibility, high wear resistance, high fracture toughness, and good compressive strength.

Bio-Glasses: An Introduction, First Edition. Edited by Julian R. Jones and Alexis G. Clare.
© 2012 John Wiley & Sons, Ltd. Published 2012 by John Wiley & Sons, Ltd.

Titanium and cobalt alloys are used in many applications where an extensive load-carrying ability is required.

Despite these interesting properties, all these high-strength materials lack osteointegration (bone bonding), which is known to be a key requirement for many biomedical applications. When implanted *in vivo*, these biomaterials, particularly alumina, show the prompt formation of a non-adherent fibrous capsule at the tissue–device interface. The presence of this fibrous capsule is not desirable for devices in which direct interaction with the surrounding tissue would be highly preferable. For this reason, modification of the surface properties by means of an appropriate coating process is of great scientific interest, as it can lead to a substantial improvement in their osteointegration. The crucial requirement to assure a chemical bond with the surrounding tissues is related to the ability of the coating to form a biologically active hydroxyapatite layer on its surface.

This requirement can be satisfied by bioactive glasses and glass-ceramics, which are widely known to promote new tissue formation on their surfaces and to encourage positive interactions between cells and implanted devices. A glass or glass-ceramic coating gives several interesting advantages with respect to the uncoated substrate [2]:

(a) It avoids substrate corrosion and degradation.
(b) It protects the surrounding tissues from adverse interactions with the degradations products of the substrate.
(c) It promotes the bioactive fixation of the implant to the living bone, inducing its osteointegration.
(d) It is easier to modify a glass coating rather than a metal implant to deliver ions such as antibacterial silver or to functionalize the glass surface with drugs or growth factors [3].

Several methods can be used to coat a substrate with a glass or glass-ceramic layer. Among them, much of the literature deals with enameling, glazing, plasma spraying, spin casting, sputtering, electrophoresis, and pulsed laser ablation. Some of these most commonly used and investigated deposition methods are described in the following sections.

8.2 ENAMELING

Enameling is a low-cost and simple method to coat metallic substrates with a glass. The term "enamel" refers to a vitreous, glass-like coating fused on to a metallic substrate. In history, enamels were first applied on precious metals (such as gold and silver), then on copper and bronze, and more recently on iron and steel. This technique was successfully proposed in the past 20 years to produce bioactive glass and glass-ceramic coatings on Co–Cr, titanium, and Ti–6Al–4V alloys, for orthopedic applications [2, 4].

Metallic substrates should initially be polished with diamond and cleaned in acetone and ethanol or chemically etched in acid solutions, in order to remove the native surface oxide layer. Usually the glass is synthesized by melting the raw materials in refractory crucibles, then the melt is quenched in air or in water (see Figure 1 in colour section and Chapters 1 and 2). The frit obtained by quenching is ground into a powder by milling and sieved below an average grain size. In order to apply the coating, the glass powders are often dispersed in a liquid medium to obtain slurries, which can be applied to the substrate by dipping, spraying, painting, and so on. Otherwise, substrates can be covered by glass powders by controlled deposition of suspensions [2]. After drying of the powders, a thermal treatment is carried out, trying to fulfill the following requirements:

(a) The firing process must be performed at an appropriate temperature that will allow a good softening and sintering of the glass powders (i.e., above the glass transition temperature, T_g) while completely avoiding any degradation of the metal. For example, in the case of titanium substrates, the firing temperature should be below the $\alpha \rightarrow \beta$ crystallographic transformation of Ti, which occurs between 885 and 950 °C for unalloyed Ti or between 955 and 1010 °C for Ti–6Al–4V, and negatively affects the mechanical properties of the metallic substrates.

(b) The firing time should be as short as possible, in order to prevent the formation of undesired reaction layers at the interface between the substrate and the coating, which can lead to poor adhesion and ease delamination of the coating.

(c) The glass should have a thermal expansion coefficient matching that of the metal that is being coated to prevent cracking as the device cools after coating.

(d) The glass coating should maintain its bioactive properties, without any contamination by metal ions diffused from the substrate, in order to form hydroxycarbonate apatite (HCA) when in contact with body fluids.

Long process times or high temperatures could cause extensive reaction between the glass and the substrate, with the formation of oxides or other products at the coating interface, and thus strongly affecting the coating adhesion. These reactions can be observed when a silica-based glass is used as the coating on Ti-based alloys or Co–Cr alloys according to [2, 5]:

$$8Ti(\text{substrate}) + 3SiO_2(\text{glass}) \rightarrow Ti_5Si_3(\text{interface}) + 3TiO_2(\text{interface})$$

$$5Ti(\text{substrate}) + 3SiO_2(\text{glass}) \rightarrow Ti_5Si_3(\text{interface}) + 3O_2$$

$$Cr(\text{substrate}) + \frac{3}{2}Na_2O(\text{glass}) \rightarrow \frac{1}{2}Cr_2O_3(\text{interface}) + 3Na(g)$$

$$Cr(\text{substrate}) + \frac{1}{2}SiO_2(\text{glass}) \rightarrow CrO(\text{interface}) + \frac{1}{2}Si(\text{interface})$$

Additionally, several reactions have been observed between Ti substrates and glasses containing P_2O_5:

$$16Ti(\text{substrate}) + 6P_2O_5(\text{glass}) \rightarrow 4Ti_4P_3(\text{interface}) + 15O_2$$

$$9Ti(\text{substrate}) + 3P_2O_5(\text{glass})$$

$$\rightarrow 2Ti_4P_3(\text{interface}) + 7O_2 + TiO(\text{interface})$$

$$17Ti(\text{substrate}) + 6P_2O_5(\text{glass})$$

$$\rightarrow 4Ti_4P_3(\text{interface}) + 14O_2 + TiO_2(\text{interface})$$

These reactions can be successfully controlled by careful optimization of the enameling treatment, and both time and temperature play equally important roles for the final coating properties. The conventional enameling theory proposes that the glass in contact with the alloy should be saturated with the lowest valence oxide of the metal, without any interfacial layers [5]. In this way, a transition region should form between the metallic bonding of the substrate and the ionocovalent bonding of the glass, providing a "continuity of electronic structure" that will result in a good bonding between the glass and the metal. Anyway, a proper tailoring of the interfacial reaction layer could improve the coating adhesion. For this purpose, Lopez-Esteban *et al.* [2] formulated a new family of glasses in the SiO_2–Na_2O–K_2O–CaO–MgO–P_2O_5 system. Glass coatings, $\sim 100\,\mu m$ thick, have been fabricated onto metallic orthopedic implants. Care has been taken to avoid excessive interface reaction, in order to prevent the formation of a thick reaction layer accompanied by bubbles in the glass and loss of adhesion.

Figure 1 Pouring melt-derived glass to make frit, which is ground and sieved to produce particulate.

Bio-Glasses: An Introduction, First Edition. Edited by Julian R. Jones and Alexis G. Clare.
© 2012 John Wiley & Sons, Ltd. Published 2012 by John Wiley & Sons, Ltd.

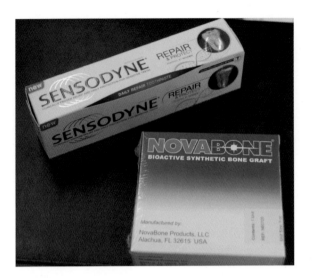

Figure 2 Two Bioglass products: Sensodyne Repair & Protect toothpaste, which contains fine Bioglass particulate to reduce sensitivity of teeth, and NovaBone, which is a particulate used as a synthetic bone graft in orthopaedics.

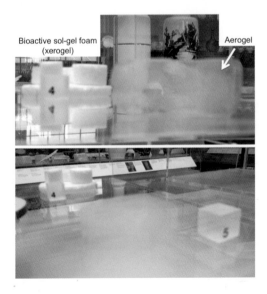

Figure 3 An aerogel photographed next to xerogel foams: (top) owing to high porosity, the aerogel is almost transparent; (bottom) depending on the angle at which it is viewed, the aerogel can appear almost invisible. Aerogels are 99% air and are made by critical point drying of sol–gel silica. The pores are nanoscale. The xerogel foam is a bioactive glass scaffold (Chapter 12) made by ambient drying of a sol–gel that has been foamed with surfactant to obtain high porosity. (This photograph was taken at the Ceramics Exhibition at the Victoria and Albert Museum, London.)

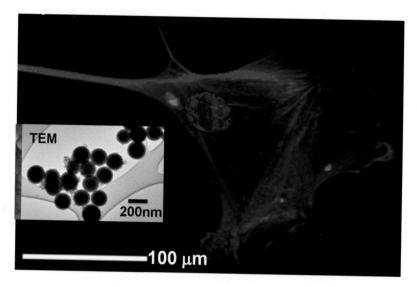

Figure 4 Sol–gel derived bioactive glass nanospheres (green and in inset) inside a bone-marrow-derived adult stem cell following *in vitro* cell culture (fluorescent staining). The cell nucleus is blue and actin filaments in the cytoplasm are red. The particles are internalised but have little effect on cell behaviour. They do not enter the nucleus (Courtesy of Olga Tsigkou and Sheyda Labbaf).

Figure 5 Glass structure at the atomic scale underpins all the properties of a glass, from degradation and bioactivity to the temperature at which it crystallises. Computer models can now be used to predict the structure of glasses and how they behave in certain conditions (Courtesy of Alastair Cormack).

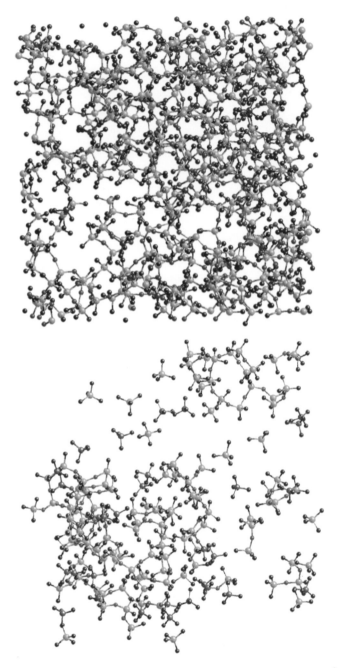

Figure 6 Advanced computer models can now predict the structure of bioactive glasses and how they behave in body fluid (Courtesy of Alastair Cormack).

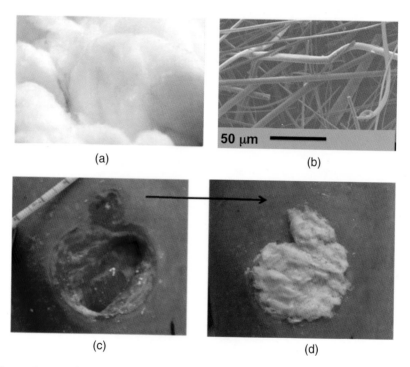

(a)

(b) 50 μm

(c)

(d)

Figure 7 (top left) A wad of bioactive borate glass nanofibres. (top right) SEM image of the fibres. (bottom left) A wound prior to dressing with the borate glass nanofibres. (bottom right) The same wound after dressing with the borate glass nanofibres (Courtesy of Steve Jung).

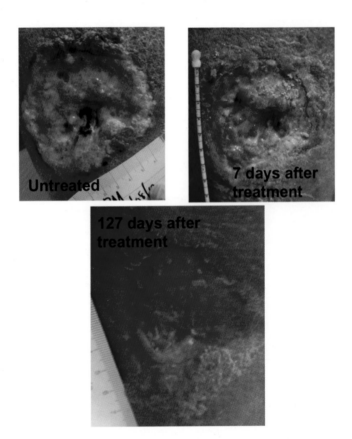

Figure 8 Large leg wound (not healing without treatment) treated with bioactive borate glass nanofibres, which healed in about four months (Courtesy of Steve Jung).

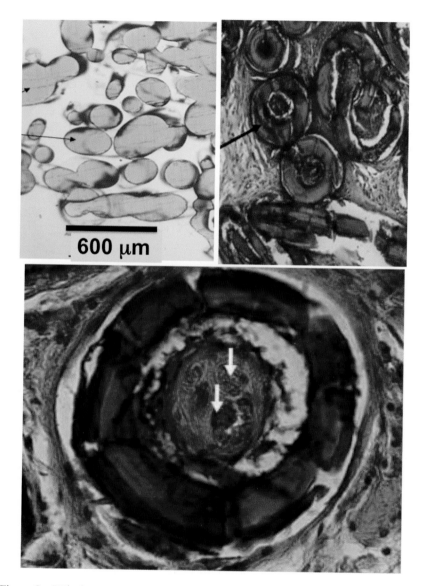

Figure 9 Histology images of bioactive borate glass fibres forming hollow channels *in vivo*. (top left) Unreacted glass fibres showing that they are originally dense glass fibres. (top right) Reacted fibres with tissue growing inside and around them after implantation. (bottom) A single hollow fibre with blood vessels inside (arrows) (Courtesy of Steve Jung).

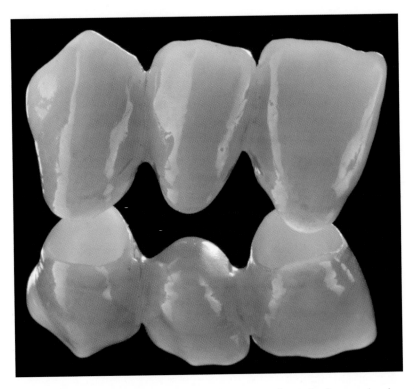

Figure 10 Long-lasting glass–ceramic teeth can now be made that can be shaped and have a colour matching natural enamel. The slight yellowing is tailored by the glass composition. Reprinted with permission from [2] Copyright (2006) W. Höland.

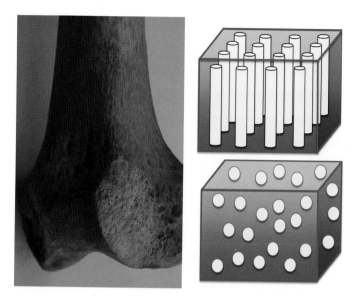

Figure 11 Photographs of a human femur. Bone is a natural composite of collagen (polymer) and bone mineral (ceramic). It also has a complex macrostructure of a dense outer layer, often supported by porous structure. This combination gives bone its great mechanical properties. The schematic shows the more simplistic structure of conventional composites. Mimicking bone is desirable . . . but a big challenge.

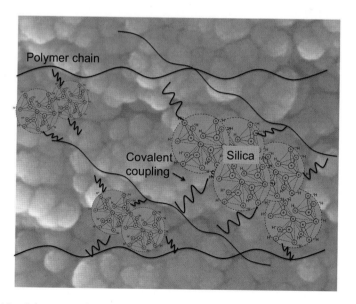

Figure 12 Schematic of the nanostructure of a sol–gel hybrid that can be bioactive, flexible and tough.

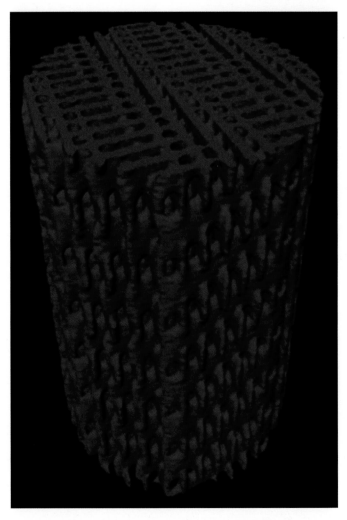

Figure 13 Strong bioactive glass scaffolds made by solid freeform fabrication using melt-derived glass (Courtesy of Eduardo Saiz and Qiang Fu).

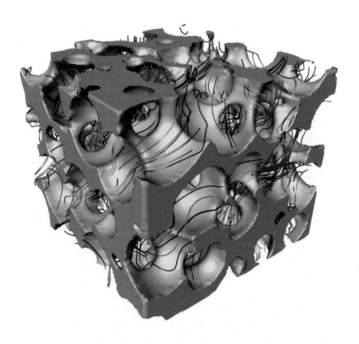

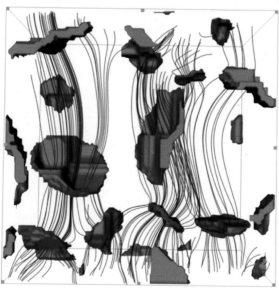

Figure 14 (top) 3D (microtomography) image of a bioactive glass scaffold (Courtesy of Sheng Yue). The lines are predicted flow paths of body fluid through the scaffold. Using image analysis techniques, the pores and interconnects between them can be quantified. (bottom, reprinted with permission from Jones *et al. Biomaterials*, 2007: **28**: 1404–1413) The interconnects of a scaffold (the scaffold itself made transparent) and predicted flow paths through them.

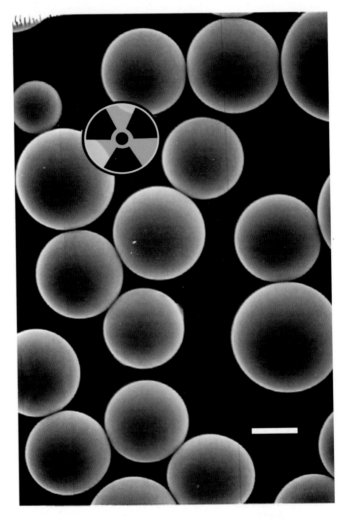

Figure 15 Radioactive glass microspheres for cancer treatment (Courtesy of Delbert Day)

Excellent adhesion to the substrates has been achieved through the formation of controlled 100–200 nm thick interfacial layers (Ti_5Si_3 on Ti-based alloys and CrO_x on Co–Cr).

Glass coatings can crack if stresses arise due to the glass shrinking at a different rate from the metal substrate as it cools. The shrinkage of the glass can be matched to that of the metal by tailoring the composition of the glass so that its thermal expansion coefficient matches that of the metal. In fact, the glass should have a slightly lower thermal expansion than the metal. This may induce only small compressive stresses, avoiding the generation of tensile thermal stresses during cooling from the processing temperature to room temperature, which may cause coating cracking or delamination during processing. Bioactive glasses are typically silica-based glasses, with silica content below 60 wt% (glasses with silica contents greater than 60 wt% are no longer bioactive; see Chapter 2). Most of these glasses have thermal expansion coefficients much higher than those of Ti alloys. Thermal expansion of the glass can be reduced by increasing the SiO_2 content, but this reduces bioactivity as well [1]. A lower thermal expansion can also be reached by a partial substitution of CaO by MgO, and of Na_2O by K_2O, matching the thermal expansion of the coating to that of Ti-based alloys. In that way, coatings with silica contents below 60 wt% that do not crack or delaminate can be successfully prepared [2]. Another method to match the thermal expansion of the coating to that of metallic substrates is to prepare multi-layer coatings. A simple method to produce bioactive glass-ceramic coatings on Ti–6Al–4V substrates by dipping and firing has been developed, in which an SiO_2–CaO–Na_2O–MgO–P_2O_5–K_2O glass is used as first layer in direct contact with the metallic substrate and an SiO_2–Al_2O_3–P_2O_5–K_2O–CaO–F^- glass-ceramic is used as outer bioactive layer [6]. The deposition of the intermediate layer was useful to obtain a good adhesion of the coating to the substrate, to minimize the reactivity between the substrate and the outer glass-ceramic coating, and thus to preserve the nature of its crystalline phases. The optimized coating method was then used to coat Ti–6Al–4V screws for dental applications.

The reactivity of glasses at the temperatures involved in the enameling process is often an issue when a bioactive coating with good mechanical and biological properties is needed. Silica-based glasses have a random network structure with many open pathways for ion diffusion. Specifically, for bioactive compositions, this property is directly related to their high ability to form HCA on their surfaces when in contact with physiological fluids. On the other hand, their open network also makes them prone to ion diffusion from the substrate toward the coating surface. This feature can severely hamper the nucleation of the calcium phosphate-rich layers and thus the formation of crystalline HCA. In fact, only a small percentage of multi-valent cations is sufficient to completely hinder the bioactivity of a glass and thus its bone bonding ability [1]. Multi-layer coatings and careful optimization of both the glass composition and the firing schedule should be good strategies to achieve this goal.

8.3 GLAZING

A thin, glossy, and glass-like coating formed on the surface of a ceramic substrate is a "glaze." The glass, synthesized by a melting and quenching route, are usually applied to the substrate by dipping, spraying, pouring, painting, and so on (as already described for enameling). An optimized thermal treatment must be carefully developed at temperatures slightly above the liquid flow temperature of the glass, obtaining layers with a thickness of about 30–150 µm. After the firing process, the coated samples should be annealed, in order to obtain tension-free glass coatings. More precisely, in order to have an amorphous layer, after the heat treatment, the coating can be simply annealed; otherwise, in order to obtain a glass-ceramic coating, the coated materials must be thermally treated with a nucleation and growth process, based on the characteristic temperatures of the pure glass (Chapter 7).

In the biomedical field, glazing can be successfully used to coat zirconia and alumina by bioactive glasses and glass-ceramics [7]. An intermediate layer often forms between the coating and the substrate – for example, between a glass coating and a zirconia substrate there is a "composite" layer made of glassy phase and zirconia particles. During the thermal treatment above its melting point, the glass diffuses within the zirconia substrate, so the zirconia granules are surrounded by a glassy matrix, leading to the formation of a "composite" layer, which assures a continuity of thermal and mechanical properties from the zirconia substrate to the glass coating.

In order to study the adherence of the coatings, the relative crack resistance can be qualitatively evaluated using indentation techniques. Indentation techniques work on the basis that an indenter (diamond tip) pressed into a harder material will travel less far into the material than it would into a softer material. Because of the complex loading

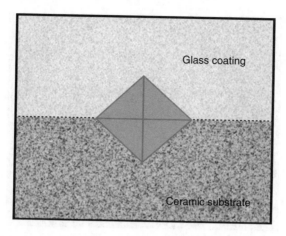

Figure 8.1 Schematic of a diamond-shaped hardness indenter with one of its diagonals on the interface of a coating and the substrate.

situations involved, a more exhaustive characterization is needed to predict the *in vivo* response of the interfaces. However, indentation testing is an attractive alternative to more costly fracture mechanics experiments for determining, in a preliminary way, the adhesion between two materials. This comparative method is based on measurements of the resistance toward propagation of a crack along an interface: cracks can be introduced by Vickers indentations and observed by scanning electron microscopy. The resistance to crack propagation provides a qualitative measurement of the strength of a brittle material. The induced radial cracks propagate in a direction parallel to the indentation diagonals and normal to the specimen surface. When the indentations are performed at the substrate–coating interface, with one of the diagonals near or just on the "border line" between the two materials (Figure 8.1), the crack propagation gives qualitative information about the fracture energy of the two joined materials and about the fracture energy of their interface. The crack path will propagate into the weaker material or it will follow the most weakly bonded interface. If the bonding at the interface is stronger than the coating, the crack would propagate through the coating.

Osteointegration of bioactive glass-coated zirconia cylinders has been evaluated in an animal model and compared to uncoated cylinders [8]. After 30 and 60 days, bone bonding was better to the coated cylinders, but after 60 days the difference was within the statistical uncertainty.

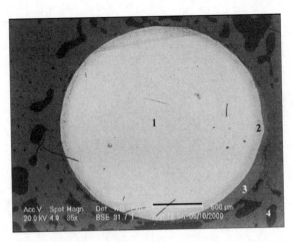

Figure 8.2   A backscatter electron micrograph of a zirconia cylinder coated with bioactive glass 30 days after surgery. Scale bar is 500 μm. (Reprinted with permission from [8]. Copyright (2002) Elsevier Ltd.)

Figure 8.2 shows a backscatter electron micrograph of a zirconia cylinder coated with bioactive glass 30 days after surgery. Four regions can be evidenced at the interface between the implant and the bone: 1, zirconia cylinder; 2, region of glass diffusion through the substrate; 3, bioactive glass; and 4, bone.

Although alumina substrates have been successfully coated by bioactive glasses [9], there are several issues that remain to be solved. As already discussed for enameling, most of the bioactive glasses for biomedical applications have a high thermal expansion coefficient, much higher than that of alumina. Alumina could be coated by glasses having a low expansion coefficient, which could be obtained by using a higher content of silica. Increased silica contents would require a higher processing temperature and, therefore, an extensive reaction between the substrate and the glass would occur. This, in turn, would lead to undesirable changes in the glass structure (crystallization), which would adversely affect its bioactivity, owing to stabilization of the glass network. Alumina may also slow bone mineralization owing to the precipitation of multi-valent ions such as hydroxides or carbonates, which are not compatible with the bone growth process.

One approach to solve this problem is based on multi-layer coatings to accommodate the challenges described above [10]. Bioactive coatings on alumina can be produced using graded structures by means of different techniques. One approach is based on multi-layer glass coatings, obtained by covering the ceramic substrate with a first layer of high-melting glass (ground coat), and one or more layers of a low-melting and less-reactive bioactive glass (cover glass).

Another method is based on glass-matrix composite coatings, obtained by adding a second phase into the bioactive glass, aiming to accommodate the mismatch in the substrate and the coating thermal expansion coefficients, acting as well as an alumina diffusion barrier. For example, aluminum oxide can diffuse from the substrate into the softened glass during the coating preparation.

8.4 PLASMA SPRAYING

Thermal spraying is a well-established technology commonly used to produce coatings for a wide variety of applications, including coating hip replacement prostheses with synthetic hydroxyapatite [11]. Thermal spray processes can be grouped into three major categories: plasma-arc spray, flame spray, and electric wire-arc spray. These three categories use different energy sources to heat a coating material (generally metals, ceramics, polymers, or their mixtures in powder form) to a molten or softened state. The heated particles are then accelerated toward a substrate. When the particles hit the substrate, a bond forms between the molten particles and the substrate.

The development of bioactive glass coatings and composites on metal substrates (Ti and its alloys) was made easier by the plasma spray technique [11, 12]. The process takes place by injecting a gas flow (argon, nitrogen, hydrogen, or helium) into a chamber where a high-temperature plasma flame is produced by means of an electric arc. The gas temperature increases up to $10\,000–30\,000\,K$. Powders of an appropriate grain size (generally around $80\,\mu m$) are injected into the chamber, rapidly heated, and accelerated through a nozzle toward the substrate. The hot material impacts on the substrate surface and rapidly cools, forming a coating. The speed of the particles may range between 100 and $350\,m/s$, and therefore the flight time is of the order of only $10^{-3}\,s$. Upon impact, the molten particles yield their thermal and kinetic energy to the substrate; they undergo deformation to a lenticular shape and solidify in less than $10^{-6}\,s$. By multiple scanning of the substrate, the particles deposit on top of one another, making up a coating of the desired thickness, usually around $100–150\,\mu m$.

Some disadvantages of this technique are its high cost, structural or morphological changes of thermodynamically unstable coating materials, and the presence of residual macro- or microporosity. However, plasma spraying offers several advantages compared to traditional enameling, such as the high deposition speed, good control of the substrate degradation, and minimum size tolerance, and it makes it possible to control the morphology, thickness, structure, and properties of the coatings by tailoring the deposition parameters. Plasma spraying is a challenging technology for the development of composite coatings. The poor mechanical properties of bioactive glasses do not allow their use in load-bearing devices. Their mechanical properties can be enhanced by reinforcing them with metallic particles (obtaining bioactive glass-matrix composite materials, i.e. "biocomposites"). These biocomposites could be prepared as bulk materials as well as coatings on tough substrates. For example, bioactive glass-matrix/Ti particle composite coatings have been successfully deposited on Ti alloy substrates by vacuum plasma spraying (VPS) [13]. In this case the bioactive coating provides both a chemical bonding to the bone and a tough behavior. In order to produce composite coatings, different methods can be used:

(a) VPS of "composite powders," that is, spraying Ti particles coated by glass, obtained by milling a sintered composite; and
(b) VPS *in situ*, that is, simply mixing glass powders and titanium particles and vacuum plasma spraying the mixture on the substrates.

Several bioactive glasses have been used as matrices for the preparation of Ti particle-reinforced composite coatings on sand-blasted Ti–6Al–4V substrates using the above-mentioned VPS methods. Generally, the composite coatings obtained by VPS of the powdered pre-sintered composites showed a better mechanical behavior with respect both to the composite coatings obtained by the *in situ* method and to the pure glass coatings. Morphological, electrochemical, and mechanical analyses showed that the sintering process is a useful step for the composite coating preparation. The best results are generally obtained by spraying "composite powders," that is, powders formed by Ti particles covered by a layer of glass obtained by milling the pre-sintered composites. In this way the softening properties of the glass matrix are fully utilized, and a better stability of the interface, both between the glass and the dispersed particles, and between the composite coatings and the Ti–6Al–4V substrate, can be guaranteed. However, the *in situ* method (i.e. the deposition of mixed glass and Ti powders) is not completely satisfactory in terms of homogeneity of the coating, particles distribution, and mechanical properties.

8.5 RADIOFREQUENCY MAGNETRON SPUTTERING DEPOSITION

In the sputtering process, surface atoms of a target material are removed and deposited on a substrate by bombarding the target with atoms of ionized gas (argon). In the radiofrequency (RF) magnetron sputtering technique, an RF power source is used. A magnet, located behind the target, enhances ionization and directs the sputtered atoms toward the substrate, where the coatings are deposited atom by atom. This technique is widely used for the deposition of insulators (i.e. silicon oxide, aluminum oxide, and other oxides), when the temperature sensitivity of the substrate precludes other techniques, or when compositional control is needed and the films or targets are insulators. Large areas can be coated uniformly without the need for substrate motion. Dense, uniform, and adherent films with similar properties as the bulk material can be obtained, and the main properties of the films can be modified by varying the deposition parameters, even if the changes induced have not yet been investigated in detail. Bioactive glass and glass-ceramic coatings from the $MgO-CaO-P_2O_5-SiO_2$ system have been successfully deposited on titanium and silicon substrates by RF magnetron sputtering [14]. The adhesion strength of the films, investigated by pull-off testing, gave very good results for amorphous films (41.1 ± 4.5 MPa), but worse for crystallized films (16.3 ± 1.9 MPa).

8.6 PULSED LASER DEPOSITION

Pulsed laser ablation provides a variety of thin coatings, from a wide range of target materials, on a wide range of substrates at room temperature. The process is described as a sequence of steps, starting with the interaction of laser radiation with a solid target, absorption of energy, localized heating of the surface, and subsequent material evaporation (ablation plume). The composition and the properties of the ablation plume may evolve through plume–laser radiation interactions and by collisions between particles in the plume itself. Finally, the plume impinges on the substrate, producing a coating.

Pulsed laser deposition (PLD) has been proposed for producing bioactive glass coatings as an alternative method to enameling and plasma spraying, because it has several processing advantages, such as the production of films from high-melting materials, stoichiometric transfer of the target composition, lack of contamination, and fine control of the film properties [15]. The introduction of a reactive gas atmosphere allows composition modulation and preserves the bonding configuration of the final coatings. The choice of the reactive atmosphere in which the plasma plume was generated affects the bonding configuration of the resulting film in terms of the presence of non-bonding oxygen groups, which are important for bioactive behavior. A disadvantage of PLD technique is that the films are not as dense as those deposited by other techniques, such as magnetron sputtering.

8.7 SUMMARY

Many different systems to produce bioactive glass coatings on metals and ceramics have been developed. Each of them has several advantages and disadvantages, beside different features, in terms of compositional homogeneity, coating thickness, ease of application, and tailoring of properties. Some of them, such as plasma spraying, can be used at an industrial scale to produce implantable devices, for both orthopedic and dental applications. A lot of data have been collected about the interface reactivity between the glass coating and the substrate, especially in the field of glass-to-metal bonding, and about the thermo-mechanical properties of the bonded materials. Several efforts are focused on the optimization of their composition to induce good bonding to the substrate as well as good biological properties. However, a word of caution is necessary: the glass coatings will degrade slowly over time (several years) and it is unclear exactly what will happen to a metal implant when that happens. This is also the case for hydroxyapatite-coated implants. Metal implants are therefore made to have a porous or pitted surface prior to coating, to try to stimulate bone ingrowth into the surface prior to the completion of degradation of the coating.

REFERENCES

[1] Hench, L.L. (1998) Bioceramics. *Journal of the American Ceramic Society*, **81**, 1705–1728.

[2] Lopez-Esteban, S., Saiz, E., Fujino, S. *et al.* (2003) Bioactive glass coatings for orthopedic metallic implants. *Journal of the European Ceramic Society*, **23**, 2921–2930.

[3] Di Nunzio, S., Vitale-Brovarone, C., Spriano, S. *et al.* (2004) Silver containing bioactive glasses prepared by molten salt ion-exchange. *Journal of the European Ceramic Society*, **24**, 2935–2942.

[4] Vitale-Brovarone, C. and Verné, E. (2005) SiO_2–CaO–K_2O coatings on alumina and Ti6Al4V substrates for biomedical applications. *Journal of Materials Science: Materials in Medicine*, **16**, 863–871.

[5] King, B.G., Tripp, H.P., and Duckworth, W.W. (1959) Nature of adherence of porcelain enamels to metals. *Journal of the American Ceramic Society*, **42**, 504–525.

[6] Verné, E., Valles, C.F., Vitale-Brovarone, C. *et al.* (2004) Double-layer glass-ceramic coatings on Ti6Al4V for dental implants. *Journal of the European Ceramic Society*, **24**, 2699–2705.

[7] Verné, E., Bosetti, M., Vitale-Brovarone, C. *et al.* (2002) Fluoroapatite glass-ceramic coatings on alumina: structural, mechanical and biological characterisation. *Biomaterials*, **23**, 3395–3403.

[8] Stanic, V., Nicoli-Aldini, N., Fini, M. *et al.* (2002) Osteointegration of bioactive glass-coated zirconia in healthy bone: an in vivo evaluation. *Biomaterials*, **23**, 3833–3841.

[9] Bosetti, M., Verné, E., Ferraris, M. *et al.* (2001) In vitro characterisation of zirconia coated by bioactive glass. *Biomaterials*, **22**, 987–994.

[10] Vitale-Brovarone, C., Verné, E., Krajewski, A., and Ravaglioli, A. (2001) Graded coatings on ceramic substrates for biomedical applications. *Journal of the European Ceramic Society*, **21**, 2855–2862.

[11] Gabbi, C., Cacchioli, A., Locardi, B., and Guadagnino, E. (1995) Bioactive glass coating – physicochemical aspects and biological findings. *Biomaterials*, **16**, 515–520.

[12] Lee, T.M., Chang, E., Wang, B.C., and Yang, C.Y. (1996) Characteristics of plasma-sprayed bioactive glass coatings on Ti–6Al–4V alloy: an in vitro study. *Surface and Coatings Technology*, **79**, 170–177.

[13] Verné, E., Ferraris, M., Ventrella, A. *et al.* (1998) Sintering and plasma spray deposition of bioactive glass-matrix composites for medical applications. *Journal of the European Ceramic Society*, **18**, 363–372.

[14] Mardare, C.C., Mardare, A.I., Fernandes, J.R.F. *et al.* (2003) Deposition of bioactive glass-ceramic thin-films by RF magnetron sputtering. *Journal of the European Ceramic Society*, **23**, 1027–1030.

[15] Borrajo, J.P., Serra, J., Liste, S. *et al.* (2005) Pulsed laser deposition of hydroxylapatite thin films on biomorphic silicon carbide ceramics. *Applied Surface Science*, **248**, 355–359.

9

Composites Containing Bioactive Glass

Aldo R. Boccaccini,[1] Julian R. Jones[2] and Qi-Zhi Chen[3]
[1]Department of Materials Science and Engineering, University of Erlangen-Nuremberg, Erlangen, Germany
[2]Department of Materials, Imperial College London, London, UK
[3]Department of Materials Engineering, Monash University, Victoria, Australia

9.1 INTRODUCTION

As we have seen in Chapter 2, bioactive glasses hold great promise in terms of stimulating bone growth, but they are brittle. They are therefore not an ideal replacement for bone, which has outstanding mechanical properties in tension, compression and even when under cyclic loading. Why is this? It is because bone has a complicated structure. The hierarchical structure of bone is a wonder of Nature. We can take inspiration from the structure and try to mimic it, but nobody has managed it yet. Bone is a natural composite of collagen (polymer) and bone mineral (ceramic). A composite material consists of two or more chemically distinct phases (e.g. metallic, ceramic or

Bio-Glasses: An Introduction, First Edition. Edited by Julian R. Jones and Alexis G. Clare.
© 2012 John Wiley & Sons, Ltd. Published 2012 by John Wiley & Sons, Ltd.

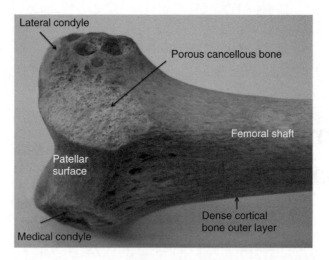

Figure 9.1   Photograph of a human femur. For a better understanding of the figure, refer to the colour section (Figure 11).

polymeric), which are separated by an interface. In bone, collagen is a triple helix of protein chains, which has high tensile and flexural strength and provides a framework for the bone structure. Bone mineral is a crystalline calcium phosphate ceramic (similar to hydroxycarbonate apatite, HCA) that contributes the stiffness and compressive strength of bone. The macrostructure of bone is also organised in a hierarchical fashion. For example, long bones like the femur have cortical bone, which is a dense structure with high mechanical strength, around the outside for stiffness. Within the long bone is trabecular bone, a network of struts (trabeculae) enclosing large voids (macropores) with 55–70% interconnected porosity, which is a supporting structure (Figure 9.1).

There is a great demand for synthetic bone substitutes that can be used in place of transplanted bone. Porous materials are also designed and manufactured to act as a *scaffold* for the growth of new bone tissue in order to restore the natural state and function, this being the fundamental aim of the tissue engineering discipline. The structure and properties of these scaffolds are pertinent to the tissue concerned and the mechanical loads it will experience *in vivo* (Chapter 12). The generic requirements for ideal synthetic bone grafts are listed in Table 9.1. An ideal synthetic bone graft would mimic the hierarchical structure of bone. However, bone structure is too complex to mimic exactly. All that can be done with current material technologies is to take inspiration

Table 9.1 Requirements for synthetic bone grafts.

1	Biocompatible
2	High porosity with an interconnected pore network to facilitate migration of cells, enabling fluid flow for nutrient supply and the removal of cellular waste products, and to permit vascular invasion
3	Suitable bioactivity to exploit the body's natural repair process, with biological response similar to that achieved by bioactive glasses
4	Biodegradable with predictable biodegradation rate matched to the formation rate of neo-tissue
5	Sufficient mechanical competence, (time-dependent) structural integrity and easy to process into 3D complex porous shapes in a controllable manner

from the composite structure and hierarchy of porosity of bone in order to develop improved synthetic composites.

Bioactive materials are an integral part of this strategy. These materials react with physiological fluids and form tenacious bonds to hard tissue through biological interdigitation of collagen fibres with HCA layers on the material surface [1]. Thus these biomaterials can be used to transfer loads to and from living bone. Bioactive glasses, as a special class of bioactive ceramics, is the subject of this volume, and detailed descriptions of the chemical composition, processing routes and properties of these materials are included in this book.

Like most ceramic materials (including bone minerals), the major disadvantage of bioactive glasses is their low fracture toughness (i.e. brittleness). Bioactive glass is therefore often used in composites combined with polymers, similar to bone minerals combined with collagen in natural bone. Both stable polymers, for example poly(methyl methacrylate) (PMMA), and biodegradable polymers, for example aliphatic polyesters, have been applied in the fabrication of biocompatible composites. However, non-degradable materials are likely eventually to be rejected by the body, owing to the cells in the body forming a fibrous capsule around the implant. Therefore, wherever possible, surgeons prefer to use biodegradable materials, as long as they complete their function before degrading and allowing natural tissue repair.

Polymer-based composites are usually made with glass or ceramic fibres dispersed in the polymer matrix (Figure 9.2) to reinforce the polymer and increase its stiffness. This is the principle used in the manufacture of lightweight high-performance components, for example, in specialist aircraft fuselages, car bodies, bicycle frames or tennis rackets. Just dispersing particles or short fibres in a matrix (Figure 9.2b) does stiffen the matrix (increase in Young's modulus) but further increase

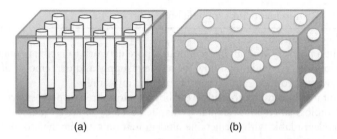

(a)   (b)

Figure 9.2   Schematics of polymer-matrix composite structures containing ceramic or glass reinforcement: (a) oriented glass-ceramic fibres and (b) glass-ceramic particles dispersed in the polymer matrix. For a better understanding of the figure, refer to the colour section (Figure 11).

in stiffness in a particular direction can be achieved by orienting the fibres (Figure 9.2a). However, when fibres are oriented, the increase in mechanical properties is only seen in the direction of the fibres – the composite is stronger in the direction of the fibres than if load is applied perpendicular to the fibre direction.

Bioactive composites containing bioactive glass within a biodegradable polymer matrix can be produced in highly porous three-dimensional (3D) structures, which are often known as tissue scaffolds because they can act as temporary templates to guide tissue growth and repair [2, 3]. Porosity of ~90% and pores of size >200 µm are desirable as well as high pore interconnectivity in order to facilitate the attachment and proliferation of cells and the ingrowth of new tissues into the network, as well as to enable mass transport of oxygen, nutrition and waste products. In general, scaffold porosity, pore morphology and pore orientation must be tailored for the particular tissue under consideration, and there is great influence of scaffold porosity and pore structure on successful bone repair.

This chapter is devoted to bioactive-glass-containing composite scaffolds for healing bone defects. We first describe the biodegradable polymers that are candidates to be combined with bioactive glasses to form porous composites for tissue engineering. Subsequently, the chapter discusses the rationale for the development of composite materials incorporating bioactive glass, describing also the technologies employed to fabricate composite scaffolds. Specific examples for bone repair are presented. The goal of this chapter is thus to allow readers to appreciate the prominent role of bioactive glasses in bioactive composite materials and to fully understand the main principles behind the development of bioactive-glass-containing composite scaffolds for bone repair.

9.2 BIODEGRADABLE POLYMERS

Bone scaffolds provide a suitable environment for new bone formation and they act as a temporary substrate to support cell attachment and subsequent bone matrix production. In order to prevent any issues associated with the long-term persistence of foreign bodies, scaffolds must be made from biodegradable materials; for example, biodegradable polymers are key components in scaffold development [4]. While the scaffold is degrading, the physical support provided by the 3D scaffold must be maintained until the new tissue formed has sufficient mechanical integrity to support itself. As the scaffold degrades, the new bone should remodel into mature bone. The degradation time profile of the scaffold must therefore be accurately controlled. Several biodegradable polymers (natural and synthetic) are being considered for bone tissue scaffolds, and detailed descriptions of the synthesis and properties of biodegradable polymers are available in the specialised literature. A brief overview of relevant polymers used in combination with bioactive glasses is presented next.

9.2.1 Natural Polymers

Natural polymers can be classified as proteins (e.g. silk, collagen, gelatin, fibrinogen, elastin, keratin, actin and myosin), polysaccharides (i.e. carbohydrates, e.g. cellulose, amylose, dextran, chitin, chitosan and glycosaminoglycans) and polynucleotides (DNA, RNA). Natural polymers exhibit similar molecular structure to the components of tissues, thus enabling easy recognition by the biological system. The obvious choice for mimicking of bone would be to use collagen. Issues related to toxicity and stimulation of inflammatory reactions, as well as lack of recognition by cells, which may be provoked by synthetic polymers, can be avoided using natural polymers. A benefit of natural polymers is that they can degrade by natural mechanisms in the body, for example, enzyme degradation, which can yield natural degradation and remodelling rates. A drawback of natural polymers is that they are difficult to produce by Nature and are difficult to produce by synthetic chemistry.

This means that natural polymers often have to be harvested from natural tissue; for example, collagen and gelatin are usually harvested from pigs. When it comes to producing medical devices and products, it can be difficult to obtain exactly the same polymer each time (e.g. inherent properties such as molecular weight), which translates into more complicated routes to regulatory approval. Patients may also object to having animal-derived products implanted on religious or cultural grounds, which would limit the market for the device. Some polymers can be grown by bacteria via biotechnology routes, but the presence of the bacteria is also a concern for regulatory bodies – and, again, reproducibility of the polymer specification is a concern. Human recombinant proteins such as collagen are being developed, but at present the process is expensive and the yields are low. The intrinsically complex structure of natural polymers, in comparison to synthetic polymers, also complicates the manufacturing techniques for the fabrication of complex structures (e.g. porous scaffolds). For example, collagen has excellent mechanical properties as it is a triple helix of three peptide chains, but its structure makes it difficult to manipulate for processing scaffolds. Collagen is used in some applications as well as gelatin (hydrolysed collagen). However, biomedical device companies prefer to make devices from synthetic polymers so that they have better control over the polymer properties, satisfying the requirements of regulatory bodies and reducing long-term risk. The problem is that there is not yet a synthetic polymer that can fully mimic the structure of polypeptides like gelatin.

9.2.2 Synthetic Polymers

Synthetic polymers represent the largest group of biodegradable polymers, exhibiting predictable and reproducible mechanical and physical properties such as tensile strength, elastic modulus and degradation rate. Unfortunately, current biodegradable synthetic polymers may degrade a bit too suddenly once the degradation process gets going (Figure 9.3).

A large proportion of the currently investigated synthetic degradable polymers for tissue engineering scaffolds are polyesters, as shown in Table 9.2. This is primarily because many biomolecules in the body and foods are esters, for example, fats and oils, which thus meet the most essential requirement on biomaterials, that is, biocompatibility.

Poly(lactic acid) (PLA), poly(glycolic acid) (PGA) and poly(lactic-*co*-glycolide) (PLGA) copolymers (Table 9.2) are among the most commonly used synthetic polymers as biomedical devices. These materials degrade on contact with water into natural products such as lactic acid and glycolic acid, and are highly biocompatible, considering that the human body already contains mechanisms for completely removing lactic and glycolic acids. PLA, PGA and their copolymers (a copolymer is when the polymer chains contain mixed species, e.g. glycolic acid and lactic acid segments) have the approval of the US Food and Drug Administration

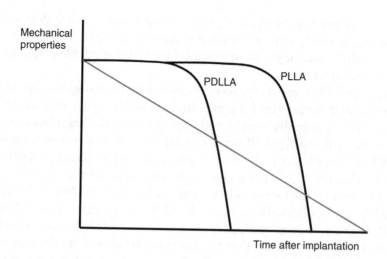

Figure 9.3 Schematic graph showing degradation profiles representative of two example polyesters (PLLA and PDLLA) that degrade by hydrolysis, compared to an ideal linear degradation profile that might be expected for polymers that degrade under enzyme degradation/remodelling.

Table 9.2 Selected properties of synthetic, biocompatible and biodegradable polymers[a] used as biomedical implants.

	P3HB	PGA	PDLLA	PLLA	PLGA	PCL
Melting temperature (°C)	170–175	225–230	not defined	173–178	not defined	58
Glass transition temperature (°C)	−4 to 10	35 to 40	55 to 60	60 to 65	45 to 55	−72
Young's modulus (GPa)	1.1–3.5	7–10	1.9–2.4	1.2–3	1.4–2.8	0.4
Water contact angle (deg.)	70–80	–	60–70	70–80	–	66
Crystallinity (%)	55–80	55–56	0	37	0	–
Degradation period (months)	>18	6–12	12–16	>24	1–12	>24

[a]P3HB, poly(3-hydroxybutyrate); PGA, poly(glycolic acid); PDLLA, poly(DL-lactic acid); PLLA, poly(L-lactic acid); PLGA, poly(lactic-co-glycolide); PCL, poly(ε-caprolactone).

for use in numerous biomedical products and devices, such as degradable sutures (dissolving stitches) and bone fixation plates. Dissolving stitches are probably one of the most common uses of degradable materials as medical devices.

PLA and PGA can be processed easily and their degradation rates, physical and mechanical properties are adjustable over a wide range by varying the molecular weight and copolymer used. Degradation is by reaction with water [5]. Water is first taken up by the polymer, causing swelling. Once water is taken up, it attacks the ester links in the polymer chains (hydrolysis), causing chain scission, reducing the chain length. PLA exists in three forms: poly(L-lactic acid) (PLLA), poly(D-lactic acid) (PDLA) and poly(DL-lactic acid) (PDLLA), a mixture of D-PLA and L-PLA. This is because the lactide monomer is chiral. PLLA can be highly crystalline, which means that the polymer chains are highly organised and tightly packed, preventing rapid water uptake. PDLLA contains a random mixture of D-lactide and L-lactide units, making it impossible to crystallise. PDLLA is therefore amorphous and the random orientation of the polymer chains allows space for water uptake. Therefore, PDLLA degrades faster than PLLA (Figure 9.3). A PLLA suture may take two years to degrade, whereas a PDLLA suture would dissolve in two months. A wide range of physical properties and degradation times can be achieved by varying the monomer ratios when using copolymers such as lactide–glycolide copolymers or hydroxybutyrate–hydroxyhexanoate copolymers. Playing with the chemistry and synthesis route can therefore give control over the degradation time. However, this does not necessarily mean control over the degradation *rate*.

Although the polyesters can be manipulated to degrade after a certain time, the degradation profile is not linear (Figure 9.3). After implantation, degradation rate is initially slow, as water is adsorbed into the polymer. The chemistry and structure of the polymer determines the rate of water diffusion. Once there is high water content inside the polymer, chain scission begins. Once the chain scission begins, the degradation process is rapid, leading to sudden loss in mechanical properties. This is because each time an ester bond is broken, a chain is cut, leaving carboxylic acid ($-COOH$) groups where the ester link used to be. Molecular weight (chain length) is reduced – on average, it will be halved with each scission. The chain scission reaction is accelerated when the pH goes away from neutral, and therefore the presence of the acidic groups accelerates degradation (autocatalysis). Once the chain length is below a critical value, the chain will leave the device. Quite quickly there will be many short chains leaving the device, which is called the 'whoosh effect'.

As this happens, mechanical properties are rapidly lost and the local pH at the site of implantation reduces rapidly because the degradation products are acids, even if they are naturally occurring acids.

The rapid release of acidic products during degradation can cause inflammatory responses, but may be mitigated through careful design of composites (Section 9.3), where the autocatalytic process can be compensated by bioactive glass components that increase pH on dissolution. In any case, assessment of the potential toxicity or inflammatory reactions as a consequence of polymer degradation is always required.

Polyhydroxyalkanoates (PHAs) belong to a type of microbial polyesters being increasingly considered for tissue engineering [6]. PHA can exist as homopolymers of hydroxyalkanoic acids, as well as copolymers of two or more hydroxyalkanoic acids. The chemically different structure of these polymers allows the development of polymer systems with different properties that affect their degradation rates in biological media as well as their mechanical properties. A major obstacle in expanding the use of polymers of the PHA group is their availability, with only two types of PHA, namely poly(3-hydroxybutyrate) (P3HB) and poly(3-hydroxybutyrate-co-hydroxyvalerate), being readily available.

Relevant properties of some of the synthetic polymers most widely used in tissue engineering are listed in Table 9.2. For applications in hard tissue repair and load-bearing sites, these polymers on their own are too weak, this being one of the reasons for their combination with bioactive glasses in composites, as discussed next.

9.3 COMPOSITE SCAFFOLDS CONTAINING BIOACTIVE GLASS

This section will discuss specifically the materials science and technology of composites based on the combination of biodegradable polymers and bioactive glass particles, added as filler or coating to the polymer matrix, for development of bone tissue scaffolds (Figure 9.4) [7].

The combination of degradable polymers and inorganic bioactive particles in the form of a scaffold is promising for bone repair, as it not only allows the combination of the 'right' biomaterials but also using a polymer matrix makes it easy to process the composite into a scaffold with a highly interconnected 3D pore network. Polymers can be easily fabricated to form complex shapes and structures, yet, in general, they lack a bioactive function (e.g. strong bonding to bone), and

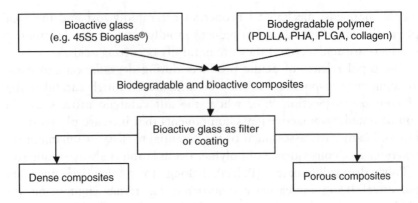

Figure 9.4 Schematic diagram showing the different combinations of biodegradable polymers and bioactive glass particles to form composite materials. (composite materials (Adapted from [7]. Copyright (2007) Woodhead Publishing.))

are too flexible and weak to meet the mechanical demands in surgery and in the physiological environment. This is where the bioactive glass reinforcement comes in.

The combination of polymers and glass particles or fibres leads to composite materials with improved mechanical properties owing to the inherent higher stiffness and strength of the inorganic material. Importantly, the addition of bioactive phases to bioresorbable polymers can alter the polymer degradation behaviour, by buffering the pH of the nearby solution and hence controlling the fast acidic degradation of the polymer, in particular in the case of the polyesters described above. Bioactive glass fillers have been shown to influence the degradation mechanism of polymers by preventing the autocatalytic effect of the acidic end groups resulting from hydrolysis of the polymer chains. Moreover, incorporation of a bioactive phase in the polymer matrix helps to absorb water owing to the internal interfaces formed between the polymer and the more hydrophilic bioactive phases, hence providing a means of controlling the degradation kinetics of the scaffolds.

The incorporation of a bioactive glass phase has an extra important function: it allows the composite to interact with the surrounding bone tissue by forming a tenacious bond via the growth of an HCA layer. However, for this to happen, processing must be optimised to ensure the glass particles or fibres are exposed at the surface of the composite, rather than being entrapped within the polymer matrix.

Two types of reinforcements are normally used for biomedical composites: fibres and particulates. The final mechanical properties of the composite are strongly affected by the volume fraction and morphology of the inclusions.

The availability of nano-sized bioactive fillers, for example, bioactive glass nanoparticles, has led to the recent development a new family of nanostructured composites for tissue engineering [6]. The higher specific surface area of nanostructured bioactive glasses allows not only for a faster release of ions from the glass and more rapid HCA layer formation but also a better distribution of particles through the polymer matrix and a better interface between the glass particles and the polymer. Nanoscale bioactive glass particles will induce nanotopographic features on scaffold surfaces, which may improve the attachment of osteoblasts (bone-forming cells), which are known to attach well to surfaces with nanoscale roughness. Bone cells are not keen on attaching directly to hydrophobic polymers like PLA, so good distribution of bioactive phases on the surface of the scaffold is paramount to the success of the composites.

The major factor affecting the mechanical properties and structural integrity of scaffolds is, however, their porosity, for example, pore volume, size, shape, pore orientation and pore interconnectivity. Hence it is very important to consider suitable fabrication technologies for production of 3D scaffolds with controlled porosity.

9.4 PROCESSING TECHNOLOGIES FOR POROUS BIOACTIVE COMPOSITES

Porous composite scaffolds have been prepared by numerous techniques, including thermally induced phase separation (TIPS), compression moulding and particulate leaching, gas foaming, and sintering of composite microspheres. Table 9.3 provides a list of typical processing techniques developed for production of highly porous bioactive composite scaffolds, including their relative advantages and disadvantages. The main techniques considered for production of bioactive-glass-containing composites are discussed in this section.

The most common method for making porous materials is the space holder or porogen leaching technique, where sacrificial particles are used as a pore template. In the case of polymer or composites, the particles (or spheres) are usually removed by washing through with a solvent that removes the template particles without affecting the polymer matrix, for example, salt crystals in a polylactide matrix. In this example, the polylactide would be dissolved in its solvent and salt particles would

Table 9.3 Fabrication routes for highly porous bioactive composites, their advantages and disadvantages. (Reprinted with permission from [8]. Copyright (2005) Expert Reviews Ltd.)

Fabrication route	Advantages	Disadvantages
Thermally induced phase separation (TIPS)	High porosities (~95%)	Long time to sublime solvent (48 hours)
	Highly interconnected pore structures	Shrinkage issues
	Anisotropic and tubular pores possible	Small-scale production
	Control of structure and pore size by varying preparation conditions	Use of organic solvents
Compression moulding/particulate (porogen) leaching	Controlled porosity	Poor interconnectivity especially at low porosities
	Graded porosity structures possible	Difficult to generate large structures (over 3 mm thick)
	No organic solvents	Not all particulates leached
Solvent casting/particulate (porogen) leaching	Controlled porosity	Structures generally isotropic
	Controlled interconnectivity (if particulates are sintered)	Use of organic solvents
Microsphere sintering	Graded porosity structures possible	Interconnectivity is an issue
	Controlled porosity	Use of organic solvents
	Can be fabricated into complex shapes	
Solid freeform fabrication (SFF)	Porous structure can be tailored to host tissue	Resolution needs to be improved to the micro-scale
	Protein and cell encapsulation possible	Some methods use organic solvents
	Good interface with medical imaging	

be added. After the lactide's solvent is removed, the polylactide will be a solid matrix. The salt particles can then be washed out with water. However, little work has been done on producing bioactive glass-polymer scaffolds using particulate (porogen) leaching. A problem with this technique is achieving adequate pore interconnectivity at low porogen (salt/sucrose) loadings, as many of the porogen particles may

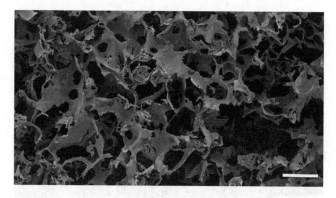

Figure 9.5 Scanning electron micrograph of a P3HB scaffold fabricated by the sugar template method. Scale bar 500 μm (Reprinted with permission from [9]. Copyright (2010) Elsevier.)

remain trapped. An alternative route has been proposed using sugar cubes as template for creation of the interconnected 3D pore structure. For example, P3HB–Bioglass® composite scaffolds have been developed by this method recently. These scaffolds exhibit a porous structure similar to the one shown in Figure 9.5 which corresponds to the neat P3HB scaffold [9]. Alternative matrices based on natural polymers, for example, starch-based polymers, have also been investigated for biodegradable polymer–Bioglass composites.

9.4.1 Thermally Induced Phase Separation

Polyester matrix composites containing bioactive glass have been developed by the thermally induced phase separation (TIPS) method. These scaffolds are highly porous, with anisotropic tubular morphology and extensive pore interconnectivity (Figure 9.6) [10]. TIPS is a phase separation procedure that uses a solvent that is easy to sublime. Dioxane can used to dissolve PLA, and phase separation is induced through the addition of a small amount of water, producing a phase with high polymer content and one with low polymer content. Freezing the solvent and then subliming it out leaves a porous construct. Controlling the freezing can be used to orient the pores. The porosity of TIPS-produced foams, their pore morphology, mechanical properties, bioactivity and degradation rate can be controlled by varying the polymer concentration in solution, the volume fraction of the reinforcing phase, the quenching temperature and the polymer and solvent used. TIPS can be considered the processing method of choice if scaffolds with highly oriented porosity

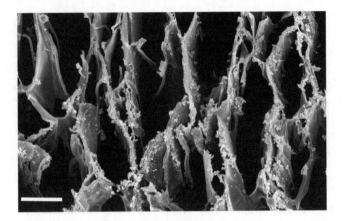

Figure 9.6 Scanning electron micrograph showing the microstructure of PDLLA-Bioglass-filled composite foam (10 wt% Bioglass) showing interconnected porosity and bioactive glass particles on the strut surfaces. Scale bar 50 μm. (Reprinted with permission from [10]. Copyright (2003) John Wiley and Sons Ltd.)

need to be fabricated. This pore structure differs considerably from that achieved by the traditional particulate salt leaching process; in that case foams exhibit a more isotropic structure with equiaxed pores but much less interconnectivity. Highly porous tubular scaffolds with oriented porosity have also been fabricated by exploiting the TIPS process. TIPS-fabricated PDLLA foams with and without Bioglass additions have been shown to exhibit mechanical anisotropy concomitant with the TIPS-induced pore architecture. For comparison, the mechanical properties of a selection of highly porous scaffolds produced by different methods are shown in Table 9.4. Inclusion of stiff inorganic bioactive phases gives slight improvement in the Young's modulus and compression strength of scaffolds. Section 9.5 will describe in detail poly(DL-lactide)–Bioglass composite scaffolds developed for bone tissue engineering by TIPS.

9.4.2 Solid Freeform Fabrication/Rapid Prototyping

Solid freeform fabrication (SFF) techniques, such as fused deposition modelling (FDM), can be used to fabricate highly reproducible scaffolds with fully interconnected porous networks. Using 3D digital data produced by computer axial tomography (CAT) scans or magnetic resonance imaging (MRI) enables accurate design of the scaffold structure. In principle, a CAT scan could be taken of healthy bone and the scan converted into a computer-aided design (CAD) file, which could be used to drive the SFF machine. The SFF machine would then

Table 9.4 Mechanical properties of several porous bioactive-glass-containing composite scaffolds.

	Elastic modulus (MPa)	Compressive strength (MPa)
PDLLA, porosity 93% (via TIPS)	0.9	0.08
Filled with 30 wt% Bioglass	1.2	0.08
PLGA (75LA : 25GA), porosity 93% (via TIPS)	0.4	0.04
Filled with 30 wt% Bioglass	0.8	0.10
PLGA (50LA : 50GA), porosity 31% (microsphere sintering)	26	0.53
Bioglass filled, porosity 43% (microsphere sintering)	51	0.42
PLLA, porosity 80% (via phase separation)	107	–
PDLLA/25 wt% Bioactive glass (porosity 77%)	145	–
PDLLA/50 wt% Bioactive glass (porosity 78%)	179	–

lay down material layer by layer and form a construct. SFF methods thereby provide enhanced control over scaffold shape, material, porosity and pore architecture, including size, geometry, orientation, branching and pore interconnectivity. However, at present, most SFF materials are 3D grid-like systems rather than exact replicas of tissue architecture (Figure 9.7). Although the scaffolds are less aesthetically pleasing than

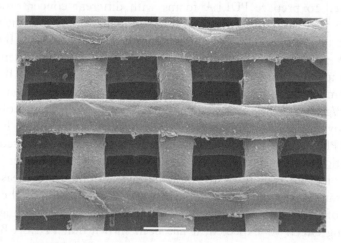

Figure 9.7 A Bioglass-polycaprolactone composite made by an extrusion-based SFF process.

a tissue replica, this may be the best method to obtain optimised mechanical properties.

9.4.3 Other Processing Routes

After the first investigations on highly porous 3D scaffolds made of bioactive-glass-filled PDLLA and PLGA, published in 2002 (see Ref. [3] for a review), an increasing amount of research has emerged on this subject and alternative methods are being proposed to fabricate composite scaffolds. For example, 3D composites of bioactive glass and degradable polymers have been produced by sintering composite PLGA–Bioglass microspheres. Sintering of the microspheres (heating the spheres until they fuse at point of contact with each other) resulted in scaffolds with interconnected porous structure. Average porosity was 40%, with pore diameters of 90 μm, and the scaffolds exhibit mechanical properties close to those of cancellous bone. The scaffolds were shown to support the adhesion, growth and mineralisation of human osteoblast-like cells *in vitro* [11].

9.5 CASE STUDY: THE PDLLA-BIOGLASS COMPOSITE SCAFFOLD SYSTEM

There has been extensive research on the PDLLA-Bioglass composite system, which will be described here in more detail as a typical case of a bioactive-glass-containing composite scaffold. The TIPS method was used to prepare PDLLA foams with different concentrations of Bioglass particles (<40 wt%) as filler [10]. Figure 9.6 shows the typical microstructure of a PDLLA-Bioglass foam (containing 10 wt% Bioglass) [10]. A porosity greater than 90% was obtained and the density of the composite foams was found to increase on addition of Bioglass; for example, the pore volume has been shown to decrease from 9.5 to 5.7 cm^3/g with addition of 40 wt% Bioglass, with little change observed in the overall pore morphology. *In vitro* studies in phosphate-buffered saline (PBS) at 37 °C showed that addition of Bioglass increased water absorption and weight loss in comparison to pure polymer foams. The molecular weight of the polymer was found to decrease at a lower rate in the composite foams, possibly as a result of the dissolution of cations (Na$^+$ and Ca^{2+}) from the Bioglass providing a pH buffering effect. As expected, it has been shown that PDLLA-Bioglass composites exhibit high bioactivity, assessed by the formation of HCA on the composite surfaces upon immersion in simulated body fluid (SBF) [3].

Owing to the potential advantages the system offers, PDLLA-Bioglass composites have been investigated in terms of their *in vitro* and *in vivo* response, as summarised elsewhere [3]. It has been shown that PDLLA-Bioglass foams support the migration, adhesion, spreading and viability of bone cells. In addition, the adhesion, growth and differentiation of human bone marrow mesenchymal stem cells (MSCs) on composites made of PDLLA foams and Bioglass particles was investigated *in vitro* and *in vivo* and the potential for *in vivo* bone formation on the composites was analysed in immuno-compromised animals. Moreover, PDLLA-Bioglass films were shown to enhance bone nodule formation and displayed enhanced alkaline phosphatase activity of primary human foetal osteoblasts in the absence of osteogenic supplements. The attachment and spreading of osteoblast cells onto PDLLA-Bioglass 3D composite foams has also been investigated. The results achieved so far have demonstrated that the regulatory role on cell differentiation and mineralisation of the Bioglass-containing PDLLA composites is likely to be a combination of both the cell–scaffold interaction (including topographic contributions) and the ionic release of Bioglass dissolution products discussed above. It is hoped that such composites can be used to stimulate repair of intervertebral discs [12].

9.6 FINAL REMARKS

The topics covered in the present chapter are intended to demonstrate how bioactive glass has a major application in the development of scaffolds for bone regeneration. This is a very active research field worldwide with bioactive glass (e.g. 45S5 Bioglass) having great biological benefits for promoting bone growth, but being brittle when used alone. Combination with biopolymers in tailored composite scaffolds can extend their application to tough load-bearing constructs. This R&D sector involves efforts from a wide range of disciplines, including cell biology, materials science, biochemistry, pharmaceutical sciences and clinical medicine. From the materials science perspective, the challenges ahead include the design and reproducible fabrication of bioactive and bioresorbable 3D scaffolds, which are able to maintain their structural integrity for predictable times, even under load-bearing conditions.

REFERENCES

[1] Hench, L.L. (1998) Bioceramics. *Journal of the American Ceramic Society*, **81**, 1705–1728.

[2] Hutmacher, D.W. (2000) Scaffolds in tissue engineering bone and cartilage. *Biomaterials*, **21**, 2529–2543.

[3] Rezwan, K., Chen, Q.Z., Blaker, J.J. and Boccaccini, A.R. (2006) Biodegradable and bioactive porous polymer/inorganic composite scaffolds for bone tissue engineering. *Biomaterials*, **27**, 3413–3431.

[4] Cooper, J.A., Lu, H.H., Ko, F.K. *et al.* (2005) Fiber-based tissue-engineered scaffold for ligament replacement: design considerations and in vitro evaluation. *Biomaterials*, **26**, 1523–1532.

[5] Hill, R.G. (2005) Biomedical polymers. In *Biomaterials, Artificial Organs and Tissue Engineering* (eds L.L. Hench and J.R. Jones). Cambridge: Woodhead, pp. 97–106.

[6] Misra, S.K., Mohn, D., Brunner, T.J. *et al.* (2008) Comparison of nanoscale and microscale bioactive glass on the properties of P(3HB)/Bioglass composites. *Biomaterials*, **29**, 1750–1761.

[7] Misra, S.K., Boccaccini, A.R. (2007) Biodegradable and bioactive polymer/ceramic composite scaffolds, Chapter 4 in "Tissue Engineering Using Ceramics and Polymers", Boccaccini, A.R., Gough, J.E., eds., (Woodhead Publishing, CRC, Cambridge, UK) pp. 72–92.

[8] Boccaccini, A.R., Blaker, J.J. (2005) Bioactive composite materials for tissue engineering scaffolds – A Review, *Expert Rev. Med. Devices* **2**, 303–317.

[9] Misra, S.K., Ansari, T.I., Valappil, S.P., Mohn, D., Philip, S.E., Starke, W.J., Roy, I., Knowles, J.C., Salih, V. and Boccaccini, A.R. (2010) Poly(3-hydroxybutyrate) multifunctional composite scaffolds for tissue engineering applications, *Biomaterials* **31**, 2806–2815.

[10] Maquet, V., Boccaccini, A.R., Pravata, L., Nothinger, I. and Jérôme, R. (2003) Preparation, Characterisation and In Vitro Degradation of Bioresorbable and Bioactive Composites Based on Bioglass®-Filled Polylactide Foams", *J. Biomed. Mat. Res.* **66A**, 335–346.

[11] Lu H.H., El-Amin S.F., Scott K.D. and Laurencin C.T. (2003) Three-dimensional, bioactive, biodegradable, polymer-bioactive glass composite scaffolds with improved mechanical properties support collagen synthesis and mineralization of human osteoblast-like cells in vitro. *J. Biomed. Mater. Res.* 64A (3), 465–474.

[12] Helen, W., Merry, C.L.R., Blaker, J.J. and Gough, J.E. (2007) Three-dimensional culture of annulus fibrosis cells within PDLLA/Bioglass composite foam scaffolds: assessment of cell attachment, proliferation and extracellular matrix production. *Biomaterials*, **28**, 2010–2020.

10

Inorganic-Organic Sol-Gel Hybrids

Yuki Shirosaki,[1] Akiyoshi Osaka,[1] Kanji Tsuru,[2] and Satoshi Hayakawa[1]

[1]Department of Bioscience and Biotechnology, Okayama University, Okayama, Japan

[2]Department of Biomaterials, Kyushu University, Fukuoka, Japan

10.1 INTRODUCTION

So far, the book has concentrated on glasses and glass-ceramics, which are very hard materials. For many applications, these materials work very well; but for other applications, softer, more flexible materials, or those with specific functional groups, are needed. For example, in bone repair applications, it may be important to gain flexibility and toughness without losing the bioactivity of the glass or glass-ceramic. Inorganic-organic hybrids are materials synthesized by combining inorganic and organic components with interactions at the molecular scale such that they are indistinguishable at the submicrometer level or above (e.g. Figure 10.1). A hybrid is very different from a composite because

Bio-Glasses: An Introduction, First Edition. Edited by Julian R. Jones and Alexis G. Clare.
© 2012 John Wiley & Sons, Ltd. Published 2012 by John Wiley & Sons, Ltd.

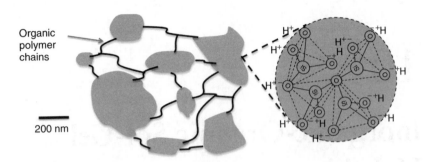

Figure 10.1   Schematic representation of the structure of a hybrid of inorganic silica and an organic polymer. Inset shows an example of the silica network.

a composite is the combination of two or more distinguishable components. Although hybrids are a relatively new type of material, they have had several names such as Ormosil, Ceramer, and hybrid. Some (incorrectly) call them nanocomposites because of the nanometer (or sub-nanometer) scale interactions between the components. At the time of writing, there are no medical devices approved for clinical use that are hybrids, but their properties mean that they are a material of the future and they have massive potential in a variety of biomedical applications.

10.2 HYBRIDS IN MEDICINE AND WHY THEY SHOULD BE SILICA-BASED

When a foreign substance is introduced into our body, cells, blood proteins, and other physiologically active molecules quickly settle on the surface to form a biological film-like layer. When ordinary artificial materials, such as metals, bio-inert ceramics, and organic polymers (synthetic), are implanted, our body produces a fibrous tissue on the surface, and expels the foreigner away. In contrast, bioactive ceramics, for example, hydroxyapatite (HA) and β-tricalcium phosphate (β-TCP), bioactive glasses (e.g. Bioglass®), and glass-ceramics (e.g. Cerabone® and Bioverit®) form a calcium phosphate layer after exposure to body fluid, instead of forming the fibrous tissue. When implanted into bone defects, they can strongly and directly bond to the surrounding bone tissue. Bioglass (45S5) has the glass composition $45SiO_2$–$24.5CaO$–$24.5Na_2O$–$6.0P_2O_5$ (wt%) (see Chapter 2). Bioverit stands for a series of glass ceramics (crystallized glass, Chapter 7) with a typical composition $45SiO_2$–$30Al_2O_3$–$12MgO$–$9(Na+K)_2O$–$4F$ (wt%). 45S5 Bioglass, the most active bioactive material, bonds to hard (bone) and soft tissue (e.g. the tympanic membrane of the ear). The dissolution products of soluble silica and calcium ions are accepted to be responsible for Bioglass having high bioactivity, as they stimulate cells to produce more bone. Calcium is common to all of the bioactive ceramics mentioned above. It would be easy to consider bioceramics as ideal materials for bone repair. However, ceramics and glasses are hard, brittle materials, with little flexibility and toughness.

Softer and flexible materials with the bioactivity of bioceramics or high affinity to either soft or hard tissue would be advantageous in clinical applications. Alternatively, bioactivity and improved stiffness can be provided to polymers through hybridization. Surgeons would like materials that could be cut with their surgical tools and pushed into a defect with their fingers. Ideally, the implant would then expand and fill the defect. They would also like the implant to share the load with the bone. This is important, as bone cells produce more natural and more healthy bone under load. Bioceramic implants are likely to fail in defects that are under cyclic loads due to their brittle nature. Implants may also be designed to be porous to act as temporary (degradable) templates (scaffolds) for three-dimensional tissue regeneration (Chapter 12). Toughness, flexibility, and bioactivity are also the desirable properties for bone scaffolds.

Suppose a large bone defect is filled with bone graft or bioactive ceramics granules are employed to fill the cavity. Fibrous tissue may invade the cavity when no membrane is applied to cover the opening. This leads to incomplete bone regeneration. Composite films from collagen or poly(lactic acid) and HA nanoparticles have been prepared for such purposes.

Medical devices that are designed to degrade are often made of synthetic polyesters, as they have a proven track record as biodegradable sutures; for example, poly(lactic acid) or its derivatives are frequently employed as pins and screws, or even as bone replacement. Although they are degradable and are not encapsulated in scar tissue, they are just as bio-inert as polyethylene when it comes to bone bonding or active stimulation of tissue repair. They do not form a hydroxycarbonate apatite (HCA) layer in body fluid and therefore do not bond with bone.

So, the obvious way to introduce flexibility and toughness into bioactive ceramics is to create a composite (Chapter 8). However, conventional composites consist of bioactive glass or ceramic fibers embedded in a polymer matrix (Figure 10.2). Mechanical properties can be engineered to be an ideal mix of those of the polymer and those of the ceramic or glass, but there are problems for biomedical applications that

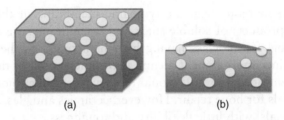

Figure 10.2 Schematics of (a) conventional polymer composite with bioactive glass or ceramic particles within a polymer matrix and (b) cell attaching to the bioactive glass or ceramic particles that protrude through the polymer matrix at the surface of the composite.

require controlled degradability or bioactivity. For example, in bone regeneration, a bioactive material is needed that can bond with the host bone and can degrade slowly as the bone regrows. One problem with a conventional composite is that the bioactive material may be covered by the polymer, masking it from the blood, cells, and host bone. This eliminates the bioactive properties. The only particles that will contact cells will be the few that poke through the polymer (Figure 10.2b), which can cause distortion of cells and reduced bone bonding. Another problem relates to degradation. If the composite is designed to degrade by using a degradable polymer matrix, it is difficult to get the polymer and ceramic or glass to degrade at the same rate. The polymer may degrade first, leaving the ceramic or glass particles without a matrix. Finally, when degradable polymers are used, mechanical properties may be lost rapidly once the polymer begins to degrade. A contributing factor is that it is difficult to get a good interface (bonding, in other words) between the ceramic or glass and the polymer. Hybrids have the potential to overcome these problems. The hope is that the fine-scale interactions between the components will mean that cells will see a hybrid implant interface as one material and the control of the chemistry will mean that a hybrid will also degrade as one material. Control of the nanostructure should also give tight control over mechanical properties and degradation rates.

Hybrids derived from atomic- or molecular-level mixtures of organic and inorganic components seem advantageous because their properties are controlled almost freely by adequate selection of the components and their mixing ratios. Hybrids can be either inorganic-rich (e.g. silica modified with some organic) or organic-rich (e.g. a polymer functionalized with inorganic regions). If the presence of soluble calcium and a silica component are important for Bioglass and Bioverit for bone bonding and osteogenic stimulation, organic materials having either Si–O bonds or Ca as their ingredients might not only exhibit good cytocompatibility but also stimulate similar reactions in the body's environment to form strong bonds with bone tissue. The hybridization not only yields flexible and bioactive materials but also causes diverse effects favorable to biomedical applications. The Si–O bonds provide a tough inorganic siloxane skeleton like that of silicone as the result of self-condensation among >Si–OH groups (the symbol > stands for additional bonds).

10.3 SELF-ASSEMBLED HYBRID FILMS AND LAYERS OF GRAFTED SILANES

Another type of hybrid is materials that are given new functionality by adding molecules to the surfaces of inorganics (Figure 10.3), for example, nanoparticles. The added functionality could be used to attach specific proteins or growth factors to an implant surface or to guide nanoparticles toward specific locations, for example, tumor targeting. Particularly, complex porous particles can contain drugs and have surface functional groups that direct them toward tumors (rather than other cells) so that they release their payload in the right place.

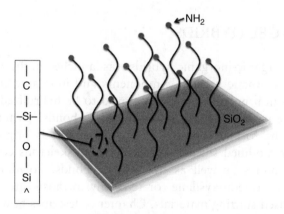

Figure 10.3 Schematic of a silica surface functionalized with aminosilanes. The $-NH_2$ groups protruding from the surface of the material change the functionality of the surface and could be used for attaching drugs or growth factors.

Bifunctional silanes with a functional group at the opposite end to the Si(OR) group can be grafted onto certain substrates by forming self-assembled monolayers (SAMs, Figure 10.3). SAMs are probably beyond the common concept of hybrids, but they can control adsorption of proteins or other physiological substances. A typical bifunctional silane for functionalization of implants or for SAM applications is amino-propyl trimethoxysilane (APTS). Figure 10.3 schematically represents their microstructure. The substrate could be a silicate glass or a silicon wafer. Hydroxyl groups on the substrate surface are condensed with the methoxy groups to form siloxane (–Si–O–Si–) bonds and an NH_2 overlayer is formed. The layer is positively charged even under neutral pH conditions, and hence the amino groups would attract negatively charged ions like phosphate ions, which means that they could induce apatite nucleation. However, the NH_2 overlayer is scarcely bioactive. Bio-inert polymers can be grafted with species (for example, a layer of 3-methacryloxypropyl trimethoxysilane) via emulsion polymerization. Substrates such as high-density polyethylene (HDPE), poly(vinyl chloride) (PVC), and polyamide (PA, nylon-6) deposit apatite in simulated body fluid (SBF), but it takes more than 14 days for HDPE while PVC and PA deposit apatite in at least 7 days. The rate of apatite formation is much lower than Bioglass (8 hours) and they cannot be said to be really bioactive. However, surface functionalization does allow control of surface chemistry of materials. This chapter will now concentrate on hybrids that have organic–inorganic interaction throughout the bulk.

10.4 SOL-GEL HYBRIDS

Synthesis of organic-inorganic hybrids was launched in the early 1980s, first among synthetic organic components and inorganic components, and then natural polymers were employed to be hybridized with inorganic counterparts involving Si–O and Ti–O bonds. Ormosils denote a series of the former type hybrids, whose name is originated from "organically modified silicates." They involve skeletons composed by >Si–O–Si< bonds as well as by >C–C< bonds. Polydimethylsilane (PDMS) and tetraethoxysilane (or tetraethyl orthosilicate, TEOS) are commonly used starting materials. Chapter 3 describes how silica glass can be synthesized by the sol-gel process. The most common procedure is to react a silicon alkoxide precursor (e.g. TEOS) with water under acidic or basic catalysis, which produces Si–OH groups. Condensation of these bonds results in oligomers and clusters. Various sizes of clusters are produced in the solution, and when those clusters grow to around

100 nm in size, they are denoted as colloid particles, and the solution is called a sol. Under acidic pH, the particles are attracted to each other and agglomerate. Further progress in hydrolysis and polycondensation causes the nanoparticles to bond to each other, and the sol sets to become a gel (a wet covalently bonded network). The time since the beginning of the hydrolysis and condensation reactions up to yielding the gel is called the gelation time or gelation period. As it is largely a room-temperature process, a polymer can be incorporated prior to gelation to create a hybrid with molecular-scale interactions between the silica and the polymer chains.

In the synthesis of a silica glass, condensation between Si–OH groups to form a silicate network represents condensation between the same inorganic species, hence it is a homo-condensation reaction.

Condensation may also take place between different kinds of M–OH groups when a different alkoxide $M(OR)_n$ is present in the reaction system, like

$$>Si-OH + HO-M< \rightarrow >Si-O-M< \qquad (10.1)$$

where M represents metals other than Si, for example, Ti or Zr, creating an inorganic hybrid. The reaction is denoted as hetero-condensation. Such a hetero-condensation reaction is difficult owing to the difference in the rate of hydrolysis reaction and the reactivity of the resulting M–OH group. For example, $Ti(OR)_4$ is more highly susceptible to hydrolysis than the corresponding $Si(OR)_4$, where R stands for an alkyl group. The hydrolysis of Si–OR is much more sluggish to yield Si–OH and, if yielded, the Si–OH group is less reactive than Ti–OH. Thus, many Ti–OH groups are produced and are condensed favorably to similar groups to form homo-clusters, Ti–O–Ti bonds, before they meet Si–OH to form Ti–O–Si bonds.

In an intermediate stage of the procedure to fabricate sol-gel hybrids, many kinds of species with different degrees in hydrolysis or condensation appear in the sol or gel systems. Owing to flexibility of the sol or gel matrix of those systems, the hybrid gels can hold molecules, large or small, like drugs, DNA or its fragments, and proteins (growth factors or enzymes), or even microorganisms. Consequently, the sol-gel processes are greatly adaptable in components or ingredients, and hence synergetic effects are expected from the hybridization among inorganic and organic species.

Organic polymers can also be introduced into the sol to improve the mechanical properties of the inorganic component, for example, improving toughness and flexibility. Organic-inorganic hybrids involving inorganic skeletons, Si–O and Ti–O, together with organic polymer skeletons are sometimes called "Ceramer" because they are ceramic-modified polymers. A typical silicate sol is made from the hydrolysis of TEOS. As condensation begins, nanoparticles form (Chapter 3). Organic polymers can then be added to the sol before the nanoparticles coalesce and gel together. This creates nanoscale interactions between the organic chains and the forming silica network. Interactions between the silica and the polymer chains can either be physical interactions such as Van der Waals forces and hydrogen bonding, or chemical covalent bonds. The latter are important if the hybrid is to have controlled degradation rates and tailored mechanical properties. Hybrids with only physical interactions are likely to dissociate rapidly in aqueous solutions, rather like sugar lumps. Covalent bonds are therefore necessary between the inorganic and organic components.

10.5 ORMOSILS

One method for obtaining covalent bonds between organic and inorganic components in a hybrid is to use an organosilane (a molecule that contains organic and silane components), for example, poly(dimethylsiloxane) (PDMS). PDMS oligomers are subject to hetero-condensation with the Si–OH groups derived from TEOS, and then the product is a typical example of an organically modified silicate, or Ormosil. Figure 10.4 represents the structure concept of Ormosils. The PDMS chains of various lengths bind the silica microblocks, derived from TEOS.

The silica blocks consist of a few silicate units, which are labeled as Q, as in glasses (Chapters 2 and 3). Each Si atom in the Q units bonds to four oxygen atoms. When an oxygen atom in a unit is bonded to a Si atom (Si–O–Si) belonging to another Q unit, it is a bridging oxygen atom. A non-bridging oxygen atom satisfies its valence by bonding to an atom other than Si, like H in Si–OH. The non-bridging oxygen can be ionically or electrostatically interacted with mono- or divalent cations: $Si-O^-\bullet Ca^{2+}$. The Q units can hold different numbers of bridging and non-bridging oxygen atoms. In contrast, the PDMS chains are composed of D units: $-Si(CH_3)_2-O-$ (D units are bifunctional) where $-O-$ represents a bridging oxygen. Note that some of the $>Si-O-Si<$ bridging bonds in the PDMS oligomer chains are decomposed by the hydrolyzing activity of HCl, added as the catalyst, during the sol-gel procedure to yield extra Si–OH groups. They are active and condensed

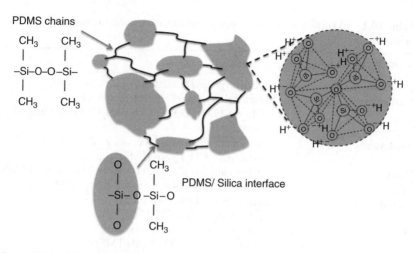

PDMS chains

CH_3 CH_3
| |
$-Si-O-O-Si-$
| |
CH_3 CH_3

O CH_3
| | PDMS/ Silica interface
$-Si-O-Si-O$
| |
O CH_3

Figure 10.4 Schematic of the nanostructure of Ormosils. Inset shows an example of the silica network.

with those on the silica blocks. The calcium ions, if added to the starting solutions, are probably present near the silica block surface or inside the blocks, being electrostatically bound to Si–O⁻, but the oxygen atoms in the PDMS blocks are less negatively charged, and hence cannot stabilize the calcium ions.

A number of silanes are applicable as the starting materials for the hybrids, including derivatives of TEOS that have some reactive organic groups substituted for ethoxy (–OEt) groups. Sometimes they involve active groups at either end. Table 10.1 lists such examples of bifunctional silanes. Among those, vinyl and methacryloxy groups are primarily subjected to homo-polymerization to form the >C–C< skeleton, while the glycidoxy group is likely to be attacked by nucleophiles (having lone-pair electrons): when, for example, the organic component contains an amino group, –NH–O–C–C– bonds will be formed. Especially reactive are mercaptans, which are easily hydrolyzed to give R′–OH together with H_2S as the by-product:

$$R'-SH + H_2O \rightarrow R'-OH + H_2S \qquad (10.2)$$

The –SH group is one of the key sources for highly biological activity of those mercaptans. In addition, some mercaptans can also attack Au to form anchoring R′–S–Au bonds, and hence they are strongly fixed on Au coatings, particles, or films. They are sometimes arranged to yield SAMs as shown in Figure 10.3.

Table 10.1 Examples of active groups or molecules for organic-inorganic hybridization.

Active groups	Molecule
Glycidoxy (epoxy)	Glycidoxypropyl trimethoxysilane (GPTMS)
Methacryl	Methacryloxypropyl trimethoxysilane, $CH_2=C(CH_3)COCO(CH_2)_3-Si(OCH_3)_3-Si(OCH_3)_3$ (γ-MPS)
(iso)Cyano	3-Isocyanatotriethoxysilane, $O=C=N-(CH_2)_3-Si(OC_2H_5)_3$
Vinyl	Vinyl trimethoxysilane, $CH_2=CH-Si(OCH_3)_3$ (VTMS)
Allyl	Allyl trimethoxysilane, $CH_2=CHCH_2-Si(OCH_3)_3$
Amino	Aminotripropyl triethoxysilane, $NH_2(CH_2)_3-Si(OCH_3)_3$ (APTS)
–SH	Mercaptopropyl trimethoxysilane, $HS(CH_2)_3-Si(OCH_3)_3$ (MPTMS)
	Mercaptopropyl triethoxysilane, $HS(CH_2)_3-Si(OC_2H_5)_3$ (MPTES)

PDMS itself can be considered a hybrid. PDMS is synthesized by condensation between organic silanes that have typical formula $R_{4-n}-SiX_n$ (X can be a halogen or another functional group that gives HO–Si and R is an organic group) and the species HO–M in Equation 10.1. One of the products of high-molecular-weight PDMS is silicone rubber. Vapor-phase condensation proceeds with the halo-silanes even under ambient conditions, since the Si–X group is very reactive. Its reactivity is illustrated by fumes evolving when $SiCl_4$ is in contact with the air: it is hydrolyzed with the humidity in the atmosphere to form nanometer-sized SiO_2 particles (fume) with HCl as the by-product. $TiCl_4$ gives TiO_2 in a similar way. This means that Si–X groups are readily hydrolyzed to yield Si–OH, and the halo-silane is susceptible to indirect condensation with the HO–M under aqueous conditions. Direct condensation also takes place between two alkoxysilanes ($R_{4-n}Si(-OR')_n$):

$$>Si-OR' + R''O-Si <\rightarrow\ > Si-O-Si < +R'-O-R'' \qquad (10.3)$$

This condensation requires higher temperature or an active catalyst. $R'_2Si(OR)_2$ is a candidate to be involved in the hydrolysis and condensation. $(CH_3)_2Si(OH)_2$ is a hydrolysis product of $(CH_3)_2SiCl_2$, whose polycondensation leads to poly(dimethylsiloxane), $[-Si(CH_3)_2-O-]$.

10.6 POLYMER CHOICE AND PROPERTY CONTROL IN HYBRIDS

For biomedical applications, when a bioactive and biodegradable hybrid is being designed to combine the bioactivity of silicate glass with the toughness of biodegradable polymers, the hybridization can yield a group of synthetic materials appropriate for a wide range of biomedical applications like repair of soft or hard tissue, protecting membranes, and scaffolds, to name but a few.

When a high proportion of bioactive silica is required, hybrids are commonly prepared via a sol-gel route. The starting reagents are dissolved in a solvent together with some catalyst for hydrolysis and condensation to yield precursor colloidal solution. The system then goes through the sol state (colloidal solution) before gelation. Porous networks can also be introduced into the hybrid and the porous structure will be tuned so as to achieve optimum cell attachment and growth (Chapter 12). The constituents of the hybrids primarily control their properties. Wide ranges of organic polymers, regardless of being naturally or synthetically originated, are applicable as well as the inorganic metal–oxygen fragments (e.g. Si–O or Ti–O) and cationic species. Hence, it is possible to fabricate an enormous variety of hybrids. Yet other factors will tune the hybrid properties, a few examples of which are listed in Table 10.2. Only small modifications of the starting agents and their mixing ratios will drastically change the mechanical and chemical properties.

Polymer choice is governed by a number of factors. The hybrid should degrade at a controllable and preferably linear rate. Then, in the course of degradation, silicate species like SiO_4^{4-} will be released from the hybrid into the body fluid and hence stimulate representation of some genes or cell growth in the nearby soft or hard tissue. Another advantage

Table 10.2 Some factors to control the property of hybrids.

Factors	Notes
Constituents	Starting materials, their reactivity
Ingredients	Catalysts, chelating agents
Composition	Mixing ratios of the constituents
Polymerization	Homo- or hetero-polymerization, sometimes controlled by catalysts or coupling/chelating agents
Microstructures	Porosity, pore size, cluster size

of hybrids over composites is that the mechanical properties should decrease in line with the degradation rate.

So, the first criterion is that the polymer should be biodegradable. The first place to look for candidate polymers would be commercially available biomedical polymers, but there are few synthetic biodegradable polymers that have been approved for clinical use. They are mainly based on polyesters such as the lactides and glycolides and their copolymers. The problem with these polymers is that, once they start to degrade, the degradation rate is rapid and mechanical properties are lost rapidly. The reason for this is the mechanism of degradation, which is chain scission by hydrolysis (the ester linkage is broken by reaction with water). The problem is that the chain scission is catalyzed by the degradation products (carboxylic acid groups).

The next criteria are these: Can covalent bonds be formed between the polymer and the silica? For sol-gel hybrids, can the polymer be incorporated into the sol, that is, is it soluble?

For tissue repair, a common strategy is to look at the natural tissue for inspiration. Most connective tissues contain collagen, and therefore collagen would be an obvious candidate, but collagen has very low solubility and is difficult to process. However, gelatin is hydrolyzed collagen, and this and other natural polymers can be used, for example, polysaccharides such as chitosan.

There are essentially two methods for obtaining covalent bonds between silica and natural polymers: using a siloxane that can introduce silica to a polymer, or using a siloxane coupling agent to bridge between a silicate network and the polymer. An example of such as siloxane is 3-glycidoxypropyl trimethoxysilane (GPTMS). A GPTMS molecule contains a glycidoxy group (epoxy ring) at one end and three methoxysilane bonds at the other (Figure 10.5). The epoxy ring can be opened by

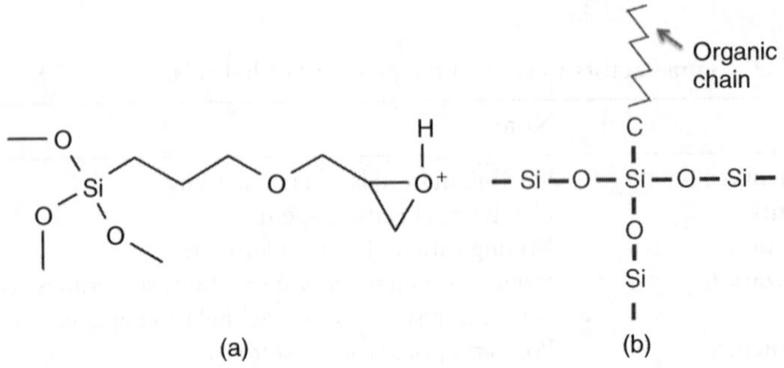

Figure 10.5   (a) The 3-glycidoxypropyl trimethoxysilane (GPTMS) molecule and (b) the T^3 unit.

nucleophilic attack and condensation can then occur with the $-NH_2$ unit of an amino acid residue on a polypeptide chain, that is, the GPTMS molecule is grafted to a gelatin molecule. The trimethoxysilane group at the other end yields $-Si-O-Si$ bridges or the siloxane units due to hydrolysis and condensation, similar to the reactions in the TEOS-based sol-gel process. Silica–organic hybrids with covalent bonds between the components consists of T-structure units ($R-Si(-O-)_3$), where T stands for trifunctional, and $-O-$ stands for a bridging oxygen atom that is shared by two Si atoms forming a bridging bond $-Si-O-Si-$. Figure 10.5(b) shows a T^3 unit with an $Si-C$ bond and three $-Si-O-Si-$ bridging oxygen bonds.

10.6.1 Silica/Gelatin

A series of gelatin–siloxane hybrids have been prepared from gelatin and GPTMS, where the epoxy ring on the GPTMS molecules open and the GPTMS is grafted onto the gelatin. Meanwhile, the trimethoxysilane group at the other end hydrolyzes, and condensation of the resulting $Si-OH$ bonds occurs to yield $-Si-O-Si$ bridges to other GPTMS bonds. In consequence, cross-links of the gelatin chains are formed. Mechanical and viscoelastic properties naturally depend on the cross-link density. For example, consider the glass transition temperature (T_g) at which the structure of a substance changes from more rigid to more flexible and fluid. Hybrids are hard and stiff below T_g, while they are soft and flexible above T_g. The T_g for the gelatin-GPTMS hybrids increases with increase in the GPTMS content, as expected from the cross-linking effects of GPTMS. When calcium is introduced by adding calcium nitrate, the presence of the calcium decreases T_g and hence gives flexibility. The effect of the calcium ions is interpreted in terms of the interaction among the gelatin-composing chains. Gelatin consists of triple helical strands, and these are uncoiled into random coils by the Ca^{2+} ions, and such a change weakened the intermolecular or intramolecular bonding among the gelatin chains with which the helix structure is stabilized.

Gelatin functionalized (grafted) with GPTMS can also be incorporated into the TEOS-based sol-gel process. The GPTMS is used to form covalent links between interpenetrating networks of gelatin and silica. The gelatin is first reacted with some GPTMS and then this hybrid is added into the early stage of the sol-gel process so that the trimethoxysilane groups on the GPTMS hydrolyze and condensation occurs between the hydrolyzed GPTMS and the nanoparticles in the colloidal sol (Figure 10.6 and see Figure 12 in colour section).

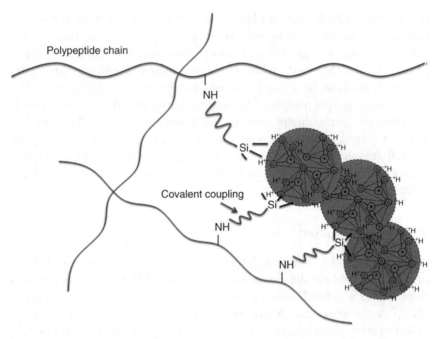

Figure 10.6 Schematic of polypeptide–silica hybrids, for example, gelatin–silica, where GPTMS is used to couple the polypeptide to the silica network formed by a TEOS-based sol-gel process.

Using TEOS to produce a silica network and GPTMS as the coupling agent allows control of the degree of coupling independent of the total inorganic-to-organic ratio, which provides a greater ability to control properties but also creates a complex system. By using this method, properties can be tailored such that the hybrid is as flexible as a polymer or as stiff as a glass, or anywhere in between. The hybrids also degrade congruently (silica is lost at a similar rate to the polymer). Poly(glutamic acid) has also been used as a simpler polypeptide in similar systems.

10.7 MAINTAINING BIOACTIVITY IN SOL-GEL HYBRIDS

In terms of a hybrid for bone regeneration, there is little point creating a hybrid if it is not bioactive. A key part of the bone bonding mechanism is spontaneous deposition of a layer of HCA (§§10.2) on the surface when the materials contact with blood plasma. The mechanism of HCA layer formation is described in Chapter 2. It is important to remember that the mechanism involves ion exchange (e.g. Ca^{2+} ions leave the glass, leaving Si–OH). The negative charge on the surface attracts Ca^{2+} and PO_4^{3-} ions. The >Si–OH and >Si–O$^-$ form a hydrated silica gel layer, in which the >Si–O$^-$ content depends on the pH of the surrounding plasma. (The symbol > stands for additional Si–O bonds to accommodate the quadruple valence state of Si.) Since the point of zero charge (pzc) of soluble silica (SiO_2) is ~pH 2, most of the silanol groups are actually present as >Si–O$^-$ groups under the body's environment at pH 7.4. Bioactive glasses can be made by the melt or sol-gel routes, and sol-gel glasses form HCA layers more rapidly than melt-derived glasses of similar nominal compositions, as they have higher surface area and lower network connectivity.

Although original bioactive glasses contained phosphate, it is not necessary for a glass to contain phosphate to be bioactive, as the blood plasma can supply phosphate (and additional calcium) ions since it is supersaturated with Ca^{2+} and PO_4^{3-} ions. An increase in the calcium ions and hydroxyl groups at the glass-blood interface triggers HCA nucleation and growth. Therefore, hybrids or composites involving >Si–OH and Ca^{2+} in their matrix are very much compatible with bone cells and surely deposit apatite when embedded in the body.

Cell–material interactions, for example, attachment, proliferation, and differentiation, result from a series of complicated surface phenomena, and depend on the material and its structure (the components from which the matrix is constructed and how they are integrated) and microstructure, such as porosity, surface charge, and surface roughness (topography). Its chemical properties also contribute to the interaction. Those factors control blood protein adsorption, that is, protein affinity, or dissolution of the components, which might affect gene expression. Adjusting the content of the silanol groups as well as the microstructure of the matrix will modify the compatibility of the hybrids with fibroblasts or nerve cells as well as osteoblasts.

10.7.1 Calcium Incorporation in Sol-Gel Hybrids

So far the hybrids discussed have largely been silica–polymer systems. However, they are not necessarily bioactive. Calcium nitrate and calcium chloride have been used in some hybrids, but the problem is that the calcium is not incorporated into the silicate network during the low-temperature processes used. Hybrids are not usually heated much above 100 °C, as higher temperatures would damage the polymer, but calcium is not incorporated into the silicate network until a temperature of

400 °C is reached, when it diffuses into the silica and disrupts the network. At lower temperatures, the calcium is likely to be precipitated on the surface of the silica or chelated on carboxylic acid groups of the polymers (if present).

10.7.2 Calcium-Containing Ormosils

Calcium-containing Ormosils are the first polymers or hybrids for which apatite formation has been confirmed since the invention of the bioactive silicate glasses. They release calcium ions into the surrounding body fluid and favor apatite precipitation. When an Ormosil-type hybrid deposits apatite in SBF, they deposit apatite within only one day. In contrast, the Ca-free Ormosils with similar compositions do not deposit an apatite layer after being soaked in SBF for 30 days, even though Fourier transform infrared (FTIR) spectrophotometry indicates that phosphate ions are attached on the Ormosil sample surface. It is perplexing that the attached phosphate should attract the calcium ions in the surrounding solution, but still those two kinds of ions would not form HCA. The HCA formation mechanism on Ca-containing Ormosils is similar to that on glasses: $Si-O-Ca^{2+}$ bonds are hydrolyzed in SBF, yielding $Si-OH$ groups and releasing the calcium ions, while hydroxide ions that increase pH are in the vicinity of the hybrid surface. Some of those $-SiOH$ groups are likely to give $-Si-O^-$ due to the hydroxyl groups, and they form the hydrated silica layer to provide the heterogeneous nucleation sites for apatite. It is strange, however, that hydrated silica does not always induce apatite nucleation, and that the active ones are either on the material surface due to the hydrolysis reaction with the body fluid or highly porous pure silica gel. Ca-free silica gels derived through a conventional sol-gel route cannot deposit HCA. Therefore, the formation of the $-Si-OH$ groups is not the only factor for controlling the apatite deposition; the way in which those silanol groups are arranged is important, that is, a specific structure is essential.

10.7.3 Ormotites

Ti–O can substitute Si–O in Ormosils, and the resultant material might be called Ormotite (organically modified titanate) in analogy to Ormosils. Some hybrids in the system $CaO-SiO_2-TiO_2-PDMS$ are similar in mechanical characteristics such as stress–strain behavior and yield strength to those of human spongy bone (cancellous bone) when their compositions and preparation conditions are optimized. Those hybrids show rather weak bioactivity, as it took three days in SBF before depositing apatite, while the bioactive Ca-Ormosils needed only a day or so. Other hybrids in the system $CaO-SiO_2-TiO_2$ have also been prepared where Ti–O–Si substitutes a part of their siloxane skeletons. They are bioactive enough to deposit apatite within one day in SBF. Cytocompatibility and blood compatibility of those Ormosils and Ormotites have been examined. A few papers have been listed in the Further Reading at the end of this chapter.

10.7.4 Hybrids From Vinylsilanes or Other Bifunctional Silanes

The silanes listed in Table 10.2 or any other silanes with a functional group at the end opposite to the $-Si(OR)_3$ end also give inorganic-organic hybrids. When vinylsilanes are polymerized, they yield polyethylene-type C–C skeletons:

$$R-CH=CH_2 + CH_2=CH-R'' \rightarrow CH(R)-CH_2-CH_2-CH(R')- \quad (10.4)$$

The Si–OH groups from the hydrolysis of the alkoxysilane groups appear randomly on the C–C skeletons. Those Si–OH groups are polymerized to yield a silica-type –Si–O–Si– skeleton. Therefore, such two active sites for polymerization can thus provide C–C, C–N (cyanosilanes), or Si–O–Si skeletons. They offer additional freedom to design a wider range of hybrids and controlling the resultant hybrids.

The vinylsilane-based hybrids deposit apatite under body conditions when they involve calcium ions. For example, a series of bulk hybrids derived from the precursor system $VTMS-Ca(CH_3COO)_2-C_2H_5OH-H_2O$ (VTMS = vinyl trimethoxysilane) deposit apatite in SBF within one day. However, even if they involve calcium ions, those gel films have an inferior ability to deposit apatite in SBF, that is, they have longer induction times. This happens partly because the gel films involved a smaller number of >Si–OH groups on their surface that would serve as nucleation sites for apatite by attracting the relevant ions, and partly because fewer calcium ions are present in the top surface. The calcium deficiency would be interpreted as a result of the following: the drying that occurs during the sol film coating process rapidly shrinks the top layer, drives down the calcium ions, and closes the paths through which they diffuse up to the top when the gel is soaked in SBF. This interpretation is supported by the shorter induction time for the samples whose surface is scarred with a knife. Thus, care should be taken on drying to increase the concentration of –SiOH groups by well hydrolyzing the –Si (OR)$_3$ groups and to secure a homogeneous distribution of calcium ions in the surface region.

10.8 SUMMARY AND OUTLOOK

The development of inorganic-organic hybrids has led to the production of new materials with unique combinations of properties, such as toughness, controlled degradation, or chemical functionality. However, synthesis of the hybrids is a complex process, and optimizing the materials is difficult with so many variables. Their potential is massive but development will take time. As they are new materials, translation to clinical products is likely also to take time and to be expensive. Upscaling laboratory-scale synthesis to mass production is also a great challenge for chemical engineers. Having said that, hybrids may be the answer to mimicking the hierarchical structure and properties of natural tissues and producing materials that fit the criteria stipulated by surgeons for their ideal implant materials.

FURTHER READING

Biodegradation of Polymers

Tabata, Y. and Ikada, Y. (1999) Vascularization effect of basic fibroblast growth factor released from gelatin hydrogels with different biodegradabilities. *Biomaterials*, **20**, 2169–2175.

Zhang, Q.Q., Liu, L.R., Ren, L., and Wang, F.J. (1997) Preparation and characterization of collagen–chitosan composites. *Journal of Applied Polymer Science*, **64**, 2127–2130.

Basic Concepts of Organic-Inorganic Hybrids

Glaser, R.H. and Wilkes, G.L. (1989) Solid-state Si-29 NMR of TEOS-based multifunctional sol-gel materials. *Journal of Non-Crystalline Solids*, **113**, 73–87.

Hu, Y. and Mackenzie, J.D. (1992) Rubber-like elasticity of organically modified silicates. *Journal of Materials Science*, **27**, 4415–4420.

Huang, H.-H., Orler, B., and Wilkes, G.L. (1985) Ceramers – hybrid materials incorporating polymeric oligomeric species with inorganic glasses by a sol-gel process. 2. Effect of acid content on the final properties. *Polymer Bulletin*, **14**, 557–564.

Iwamoto, T. and Mackenzie, J.D. (1995) Ormosil coatings of high hardness. *Journal of Materials Science*, **30**, 2566–2570.

Mackenzie, J.D., Chung, Y.J., and Hu, Y. (1992) Rubbery ormosils and their applications. *Journal of Non-Crystalline Solids*, **147**, 271–279.

Mackenzie, J.D., Huang, Q-X., and Iwamoto, T. (1996) Mechanical properties of ormosils. *Journal of Sol-Gel Science and Technology*, 7, 151–161.

Novak, B. (2003) Hybrid nanocomposite materials – between inorganic glasses and organic polymers. *Advanced Materials*, 5, 422–433.

Philipp, G. and Schmidt, H. (1984) New materials for contact lenses prepared from Si– and Ti-alkoxides by the sol-gel process. *Journal of Non-Crystalline Solids*, 63, 283–292.

Schmidt, H. (1985) New type of non-crystalline solids between inorganic and organic materials. *Journal of Non-Crystalline Solids*, 73, 681–691.

Schmidt, H. (1988) Chemistry of material preparation by the sol-gel process. *Journal of Non-Crystalline Solids*, 100, 51–64.

Valliant, E.M. and Jones, J.R. (2010) Softening bioactive glass for bone regeneration: sol-gel hybrid materials. *Soft Matter*, 7, 5083–5095.

Nanoparticles With Functional Groups for Drug Delivery

Ashley, C.E., Carnes, E.C., Phillips, G.K. *et al.* (2011) The targeted delivery of multicomponent cargos to cancer cells by nanoporous particle-supported lipid bilayers. *Nature Materials*, 10, 389–397.

Sol-Gel Synthesis of Materials

Brinker, C.J. and G.W. Scherer (1990) *Sol-Gel Science: The Physics and chemistry of Sol-Gel Processing*. Academic Press, San Diego, CA: USA.

Brinker, C.J. *et al.* (eds) (1986–1998) *Better Ceramics Through Chemistry, II–VIII*, Materials Research Society Symposium Proceedings. Pittsburgh, PA: Materials Research Society.

Cho, S.-B., Nakanishi, K., Kokubo, T. *et al.* (1996) Apatite formation on silica gel in simulated body fluid: its dependence on structures of silica gels prepared in different media. *Journal of Biomedical Materials Research*, 33, 145–151.

Bioactive Ormosils and Ormotite

Chen, Q., Miyata, N., Kokubo, T., and Nakamura, T. (2001) Effect of heat treatment on bioactivity and mechanical properties of PDMS-modified $CaO–SiO_2–TiO_2$ hybrids via sol-gel process. *Journal of Materials Science: Materials in Medicine*, 12, 515–522.

Tsuru, K., Aburatani, Y., Yabuta, T. *et al.* (2001) Synthesis and in vitro behavior of organically modified silicate containing Ca ions. *Journal of Sol-Gel Science and Technology*, 21, 89–96.

Tsuru, K., Ohtsuki, C., Osaka, A. *et al.* (1997) Bioactivity of sol-gel derived organically modified silicates. 1. In vitro examination. *Journal of Materials Science: Materials in Medicine*, 8, 157–161.

Gelatin–Siloxane and Chitosan–Siloxane Hybrids

Ren, L., Tsuru, K., Hayakawa, S., and Osaka, A. (2001a) Synthesis and characterization of gelatin–siloxane hybrids derived through sol-gel procedure. *Journal of Sol-Gel Science and Technology*, 21, 115–121.

Ren, L., Tsuru, K., Hayakawa, S., and Osaka, A. (2001b) Incorporation of Ca^{2+} ions in gelatin–siloxane hybrids through a sol-gel process. *Journal of the Ceramics Society of Japan*, **109**, 406–411.

Ren, L., Tsuru, K., Hayakawa, S., and Osaka, A. (2001c) Sol-gel preparation and in vitro deposition of apatite on porous gelatin–siloxane hybrids. *Journal of Non-Crystalline Solids*, **285**, 116–122.

Shirosaki, Y., Tsuru, K., Hayakawa, S. *et al.* (2005) In vitro cytocompatibility of MG63 cells on chitosan-organosiloxane hybrid membranes. *Biomaterials*, **26**, 485–493.

Sol-Gel Silica-Polypeptide Hybrids

Mahony, O., Tsigkou, O., Ionescu, C. *et al.* (2010) Silica-gelatin hybrids with tailorable degradation and mechanical properties for tissue regeneration. *Advanced Functional Materials*, **20**, 3835–3845.

Poologasundarampillai, G., Ionescu, C., Tsigkou, O. *et al.* (2010) Synthesis of bioactive class II poly(γ-glutamic acid)/silica hybrids for bone tissue regeneration. *Journal of Materials Chemistry*, **40**, 8952–8961.

Blood Compatibility of Some Organic–Inorganic Hybrids and Cell Culture

Deguchi, K., Tsuru, K., Hayashi, T. *et al.* (2006) Implantation of a new porous gelatin–siloxane hybrid into a brain lesion as a potential scaffold for tissue regeneration. *Journal of Cerebral Blood Flow and Metabolism*, **26**, 1263–1273.

Zhang, H-Z., Hayashi, T., Tsuru, K. *et al.* (2007) Vascular endothelial growth factor promotes brain tissue regeneration with a novel biomaterial polydimethylsiloxane–tetraethoxysilane. *Brain Research*, **1132**, 29–35.

11
Dental Applications of Glasses

Leena Hupa[1] and Antti Yli-Urpo[2]
[1]Process Chemistry Centre, Åbo Akademi University, Åbo, Finland
[2]Institute of Dentistry, University of Turku, Turku, Finland

11.1 INTRODUCTION

Biomaterials have been used for centuries to replace missing teeth. More than 2000 years ago, the Etruscans replaced missing teeth with animal bones and animal teeth framed in an arch of gold. Biomaterials are used more frequently in modern dentistry than in any other field of medicine. In dentistry, all classes of biomaterials can be applied: metals, polymers, glasses, ceramics and composites. Although the utilisation of bioactive glasses has been studied in several dental applications, only a few commercial bioactive dental product applications are currently available.

Biomaterials face a challenging environment in dental applications. The biomaterial might be in contact with both hard and soft tissues in an environment – the oral cavity – where a large number of microorganisms are present. The material must therefore not support the adhesion and growth of microorganisms, yet the material needs to adhere to biological

Bio-Glasses: An Introduction, First Edition. Edited by Julian R. Jones and Alexis G. Clare.
© 2012 John Wiley & Sons, Ltd. Published 2012 by John Wiley & Sons, Ltd.

tissue. The biomaterial must be biocompatible or bioactive, non-toxic, non-allergenic and possess an aesthetic appearance. In restorative dentistry, the biomaterial needs to have high mechanical strength, stiffness and hardness. In several dental applications the biomaterial used is shaped in site by the dentist. Thus, the material used must be easy to sterilise, to apply in site and to form into any desired shape.

The subject of materials for medical applications is highly interdisciplinary – the materials scientist must collaborate with physicians. Only then can the materials scientist understand the requirements the clinical environment has for the development and manufacture of the biomaterial. It is necessary that the final product possesses all desired properties in an easily applicable product form. The most important research of the usage of melt-quenched bioactive glasses in dental applications is reviewed below.

11.2 STRUCTURE OF THE HUMAN TOOTH

In order to thoroughly understand potential dental applications of bioactive glasses, the structure of the human tooth and its anchorage to the jawbone must be understood. Figure 11.1 illustrates simplified sketches of the structures of the teeth. The root of the tooth consists of dentine and pulp surrounded by cementum. The pulp, commonly known as the nerve of the tooth, is filled with soft tissues, including blood vessels and nerve cells. The root is embedded in the jawbone, which forms the

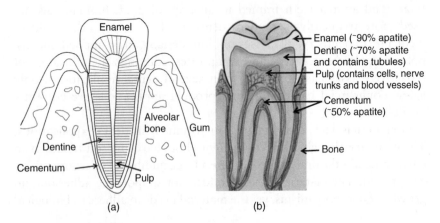

Figure 11.1 Simplified structures of teeth: (a) an incisor and its attachment to the alveolar bone, and (b) a molar.

alveolar arch. The upper alveolar jaw is called the maxilla and the lower is called the mandible. The part of the tooth visible in the mouth consists of enamel. The gum or gingiva is the soft tissue that overlays the jaws. The gum participates in attaching the tooth to the jaw.

Hydroxyapatite, $Ca_{10}(PO_4)_6(OH)_2$ (HA), is the main inorganic material in enamel, dentine, cementum and general bone tissue. Bone tissue is basically a composite made of collagen fibres and HA crystals mainly bonded with hydrogen bonds via OH groups. HA is a versatile material because it allows a high degree of replacement of its cations and anions without the collapse of the unit cell. Thus, the calcium ions in HA can be partly replaced by Na^+, K^+, Mg^{2+} or Sr^{2+}, while the phosphate can be replaced by carbonate or hydrogenphosphate, and the hydroxide ions by chloride or fluoride. The HA in living tissue always has the anions and cations substituted to some degree. This is why biological apatite is often named hydroxycarbonate apatite (HCA). The size of the biological apatite crystals is much smaller in skeletal bone tissue and in dentine apatite than in enamel apatite. The needle-shaped crystals in tooth enamel have the approximate size of 30 nm × 90 nm. In comparison, in dentine and bone, the size of the crystals is only approximately 5 nm × 30 nm. In bone tissue and in dentine, about 70 vol% consists of inorganic crystals, yet in tooth enamel the amount of crystals is 99 vol%.

11.3 GLASS BIOACTIVITY AND TEETH

The term bioactivity is usually defined as the ability of a material to bond chemically with biological tissue. Most studies of bioactive glasses in dentistry deal with the capability of the glass to support alveolar bone regeneration around the root of the tooth. Bioactive glasses have also been tested in treatments of dentine and enamel. In these applications, it is essential that the glasses partly dissolve in body fluids, thus depositing HCA crystals on the surface of the glass. Depending on the composition of the glasses, they bond with the biological apatite and guide and stimulate the growth of new biological apatite. This principle has been used to develop toothpastes that mineralise dentine to treat hypersensitivity (see later) to prevent people feeling pain when they drink hot or cold beverages.

Loss of biological apatite in alveolar bone, dentine or enamel is mostly initiated by colonisation of certain microorganisms on the tissue surfaces. The microorganisms can cause inflammatory reactions, which in turn lead to resorption of the bone, dentine or enamel. The microorganisms can also colonise on the surface of dental implants and cause bone loss, which can lead to loosening of the implant. Certain bioactive glasses have shown antimicrobial effects. The dissolution of the glasses in body fluids leads to an interfacial environment that prevents the attachment and growth of microorganisms. As bioactive glasses enhance bone regeneration and inhibit growth of various microorganisms, they have great potential in dental applications.

The tissue bonding of bioactive silicate glasses can be related to their ability to form HCA layers on their surfaces in body fluids [1]. The HCA is critical for tissue bonding – when the layer is formed, the biological mechanisms of bonding will start. Ideally, the new tissue is remodelled at a rate equal to that at which the glass dissolves.

In the mid-1980s, a systematic investigation of glasses suitable for use as bioactive coatings on metal prostheses was started [2]. One of the glasses, S53P4 glass, was adapted for further studies in the field of bone regeneration applications. Two glasses, 45S5 glass (Bioglass®) and S53P4 glass (BonAlive®), are approved by the US Food and Drug Administration (FDA) for certain clinical applications. The oxide compositions of the melt-derived glasses 45S5 and S53P4 are presented in Table 11.1.

Both glasses are available as particulates for repair of bone defects in dental applications. Bioglass is sold as PerioGlas® (NovaBone Products LLC, Alachua, Florida). Over the years, several studies of the *in vitro* and *in vivo* properties and the clinical applications of 45S5 and S53P4 glasses have been performed. However, both glasses have been tested simultaneously only in very few studies. Generally, 45S5 glass reacts more rapidly than S53P4, which is also suggested by the respective silica contents of the glasses. When cones were implanted in rat femur and soft tissue, HCA layers formed on both glasses [3]. In all implantation tests, the layer thicknesses were slightly higher for 45S5 than for S53P4. Scanning electron micrographs of the cross-sections of 45S5 and S53P4 cones after eight weeks in femur are shown in Figure 11.2. The images were taken after a push-out test of the cones. Both glasses showed

Table 11.1 Oxide compositions of 45S5 glass and S53P4 glass, in mol% (wt%).

Glass	SiO_2	Na_2O	CaO	P_2O_5
45S5	46.1 (45)	24.4 (24.5)	26.9 (24.5)	2.6 (6)
S53P4	53.8 (53)	22.7 (23)	21.8 (20)	1.7 (4)

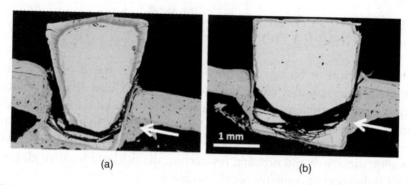

Figure 11.2 Cross-sections of cones of (a) 45S5 glass and (b) S53P4 glass after eight weeks in rat femur. The arrows point out the new bone. (Expanded view of Figure 11.2 in [3]. Copyright (2010) Society of Glass Technology.)

good contact with the bone, as indicated by the implant failure within the glass.

Glass composition has a great effect on bioactivity and processability. Chapter 2 described the effect of silica content on bioactivity and how network modifiers could be replaced with others to change properties; for example, replacing calcium oxide with strontium oxide has been found to enhance bone formation.

The number of different network modifiers can also be changed. To make porous glasses (Chapter 12), coatings (Chapter 8) or fibres from melt-derived glasses, the glasses have to be worked above their glass transition temperature. The low silica content and high lime content of the first bioactive glasses make them prone to crystallise when they are above the glass transition temperature. This is the reason why current commercial bioactive glasses are mainly particulates. Glasses containing more oxides can increase the temperature difference between the glass transition temperature and the crystallisation onset temperature, creating a wider operating window. K_2O, MgO, B_2O_3 and Al_2O_3 can be added to the glasses [2, 4, 5] to do this. The compositions of the studied glasses were chosen with the help of statistics, thus enabling the modelling of the measured properties as functions of the oxide composition of the glass. Today, models of *in vitro* and *in vivo* reactivities as well as models of the hot working properties are available [2, 4–6]. These property models can be utilised to tailor compositions for controlled reactivity in any product form desired.

11.4 BIOACTIVE GLASS IN DENTAL BONE REGENERATION

The utilisation of bioactive glasses in dentistry is based either on the ability of the glass to enhance bone, dentine and enamel apatite regeneration, or the antimicrobial effect that the dissolution of the glass in body fluids has on oral microorganisms. Most studies of bioactive glasses in dental applications have been done with 45S5 glass and S53P4 glass. Despite the active and ongoing research in the field of bioactive glasses, their utilisation in dentistry is still modest.

Dental implants require that sufficient bone tissue exists in the jaw in order to support the anchorage of the implant. After removing the tooth, the bone in the jaw underneath it starts to resorb, as it is not mechanically loaded. The thickness of the jawbone can be restored by grafting the bone with a suitable material. Usually, an autograft, that is, bone tissue taken from other sites of the same patient, is used to generate bone growth. The utilisation of autograft obviously requires additional surgery, which prolongs convalescence. The availability of autograft for large bone defects is also limited. Bioactive glasses, with their ability to guide and stimulate new bone growth, are particularly feasible materials as bone grafts.

One of the first commercial applications of bioactive glasses in dentistry was to prevent the resorption of alveolar bone in the jaw after tooth removal prior to dental implant surgery [7]. Cones of Bioglass were placed in the fresh cavities that were left after the tooth removal, and after a few weeks dentures could then successfully be implanted.

Periodontitis, an inflammatory disease caused by microorganisms in the oral cavity, has also been successfully treated with bioactive glass. In periodontitis, the supporting tissue around the teeth is infected and this infection may lead to the loosening of the soft tissue from the tooth, to bone resorption and, in cases of extensive resorption, to complete loss of the tooth. Clinical studies of filling bone defects around the root of the tooth with particles of S53P4 glass suggested that the glass is a very promising grafting material. Since then, bioactive glass particles have been used more frequently in alveolar bone healing, for example, PerioGlas (from NovaBone).

The bone in the lower jaw, the mandible, consists mainly of compact cortical bone tissue, which can be grafted with biomaterials relatively easily. The upper jaw, the maxilla, consists of porous cancellous bone tissue and resorbs rapidly in periodontitis or after removal of the tooth.

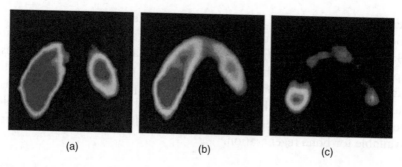

(a) (b) (c)

Figure 11.3 PET images of left and right maxillary cavities (a) 12, (b) 26 and (c) 78 weeks following maxilla grafting. The left cavity was filled with the bioactive glass S53P4 and autologous bone, while only autologous bone was placed in the right cavity. The darker grey areas indicate cell activity in the cavities due to grafting material dissolution.

The amount and quality of the bone tissue in the maxilla is rarely sufficient for dentures. Thus, grafting of the maxilla is problematic.

Bone resorption in the upper jaw is usually treated with a procedure called lifting of the maxillary sinus floor. In such treatment, bone tissue is grown partly into the sinus cavity above the maxilla. Maxilla grafting with granules of S53P4 glass together with autologous bone has been successful in facilitating the implantation of titanium roots even in the very porous upper jawbone [8]. Figure 11.3 shows positron emission tomography (PET) images of left and right maxillary cavities. The left cavity was filled with the bioactive S53P4 glass and autologous bone, while only autologous bone was placed in the right cavity. The darker grey areas indicate cell activity in the cavities due to grafting material dissolution after 12, 26 and 78 weeks, respectively. The cavity filled with bioactive glass shows more intense activity at each successive time point. The histological analyses of samples with the cancellous bone close to the cavities showed thicker bone lamellae under the cavity filled with bioactive glass and autologous bone than the samples with the cavity filled with autologous bone alone. The autologous bone resorbed from the maxilla cavity as it became surrounded by porous bone, while the mixture of bioactive glass and autologous bone reduced the amount of bone needed for augmentation of the maxillary sinus floor. The cell activity in the cavities suggests that formation of new bone will continue longer when using bioactive glass than autologous bone.

Over the years, several *in vivo* studies of bioactive glass granules have been conducted to verify the potential of bioactive glasses in four different applications: (i) as graft material to prevent loss of bone after removal of a tooth; (ii) to heal and regenerate bone defects around a tooth root and metal implant; (iii) to regenerate bone before denture placement; and (iv) to fill bone voids in implant placement. All these applications have one thing in common – the proven capability of the glass to guide and support bone growth by HA precipitation on the glass surface and the surrounding tissue. Commercial products based on melt-derived glasses, Bioglass (e.g. PerioGlas), BonAlive and StronBone™, are available for bone regeneration.

11.5 TREATMENT OF HYPERSENSITIVE TEETH

A common symptom of hypersensitivity is sharp pain when the tooth is suddenly subjected to hot, cold, bitter and sweet drinks and food. Inflammatory reactions of the gum around the teeth may lead to exposed dentine. Dentine consists of narrow $1-2\,\mu m$ long tubules that lead to the pulp of the tooth and sensitive nerve endings. Fluid flow in the opened dentine tubules is assumed to be the reason for hypersensitive teeth. Since the first results of using S53P4 glass in the treatment of hypersensitive teeth were published, several other studies have been conducted on the effect of bioactive glass and glass-ceramics to treat dentine hypersensitivity [9].

Sensodyne Repair & Protect (GlaxoSmithKline, UK) contains a fine particulate of Bioglass that reduces hypersensitivity (see Figure 2 in colour section). Trials based on brushing with NovaMin® indicate that the glass particles attach to the dentine, release ions and raise pH, causing precipitation of HCA over the tubule ends, blocking the tubules. Figure 11.4 shows dentine tubules and how the bioactive glass particles attach to the dentine and then produce HCA [10]. When the oral area is exposed to acidic environments such as carbonated drinks and fruit juices, it is likely that the HA crystals formed from the treatment with the toothpaste dissolve easily [11]. Tailoring the composition of the glass to dissolve fluorapatite crystals instead of HCA crystals has been reported to give a more chemically stable treatment against dentine hypersensitivity [11]. Similar results were observed when dentine was treated with a paste containing fine-grained S53P4 glass (diameter $\sim 20\,\mu m$).

Bioglass in toothpaste also aids in remineralisation of tooth enamel after acid leaching or bleaching treatments [12, 13]. Also, bioactive glasses can be utilised to strengthen the dentine tubules during the treatment of enamel and root caries. Stronger dentine increases the fracture strength of the restored tooth. After removing the carious tissue, the cleaned cavity is filled temporarily with a paste or suspension containing

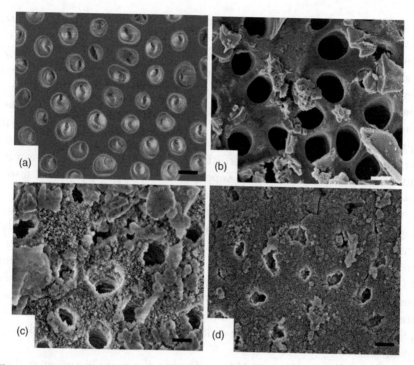

Figure 11.4 Scanning electron micrographs of (a) untreated and (b–d) treated human dentin: (b) immediately after, (c) one day after and (d) five days after application of bioactive glass in artificial saliva. (Scale bar = 1 μm.) (Adapted with permission from [10]. Copyright (2011) Professional Audience Communications Inc.)

powdered bioactive glass. The glass dissolves high concentrations of calcium and phosphate ions in close vicinity of the dentine tubules over the time needed for the ions to diffuse as deep as possible into the tubules. Then the ions nucleate, thus creating HA crystals that occlude and strengthen the tubules before the permanent filling of the cavity.

11.6 BIOACTIVE GLASS COATING ON METAL IMPLANTS

The use of glasses as solid implants and prostheses in load-bearing applications is limited by the inherent brittleness of glasses. Most dental implants are shaped like screws or cylinders made out of pure titanium or titanium alloys. In general, metal does not bond to tissue and thus metal is fixed mechanically or through bone cements to the bone. Achieving osteointegration, a stable attachment to the bone, can be improved by surface coatings that give chemical bonding to the tissue. Coatings on metal prostheses were one of the very first applications of bioactive glasses. However, using bioactive glasses as coating materials has several limitations. The glasses are brittle and do not withstand tension or bending very well. The glass might detach quite easily from the surface of the metal. The glasses are also prone to devitrify; thus, thermal treatments during coating may lead to crystallisation of the glass. Further, bioactive glasses dissolve gradually and do not provide a permanent fixation of the prosthesis. The coating might, however, give a firm initial attachment, support bone growth and give a faster osteointegration of a cementless metallic prosthesis.

Chapter 8 explains that, when enamelling metals with glassy coatings, the composition of the coating should be adjusted to match the thermal expansion coefficient of the metal in order to prevent glass from cracking and peeling off. The glass must also be designed not to crystallise during the coating process. A number of methods are utilised to manufacture thin glassy bioactive coatings on metal prostheses. These methods include, among others, liquid thermal spraying, plasma spraying, electrophoretic deposition, sol-gel coating and laser ablation.

Moritz and co-workers [14] have developed a manufacturing method that makes it possible to fabricate implants with different bioactive areas for soft tissue and bone attachments. Figure 11.5 schematically shows the soft tissue and bone attachments of a dental screw-type implant and of an ideal implant, with different areas of the bioactive coating

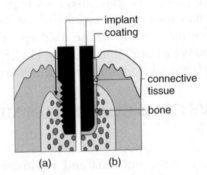

(a)   (b)

Figure 11.5 (a) An uncoated dental implant and (b) an ideal implant with areas of the coating providing local attachment to the surrounding tissue. The composition of the coating can be tailored along the implant surface. The arrow shows the resorption of bone that occurs when using screw-type implants. (Image provided courtesy of Niko Moritz. Copyright (2012) Niko Moritz.)

providing local attachment to different parts of the surrounding tissue. In the ideal implant, the coating compositions must be tailored to provide local attachment to both the bone and the gum.

The coating was manufactured by dip coating the titanium implants in a suspension of ethanol and powdered glass (<45 µm). Then the glass particles were sintered with a focused CO_2 laser beam to a coating that consists of micro-sized (diameter ∼60 µm) glass drops [14]. The glass composition used, 1-98, is optimised to show a lower tendency to crystallisation in thermal treatments than 45S5 glass and S53P4 glass. Moritz and co-workers compared the osteointegration of the bioactive glass-coated titanium implants with titanium implants in rabbit femur. After eight weeks, the implants coated with the bioactive glass were covered with significantly more bone tissue than the totally titanium implants (Figure 11.6). In the area spread 250 µm from the perimeter of the implant, significantly more bone was observed in the implants with the bioactive glass coating than in the two non-glass-coated implants. In the area spread 1 mm from the perimeter of the implant, no statistically significant difference between the materials was detected, although slightly more bone had formed around the implants coated with the bioactive glass. The results all indicate that bioactive glass coatings on titanium implants enhance the initial osteointegration.

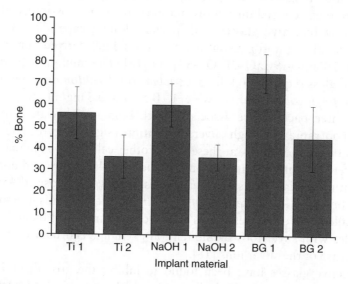

Figure 11.6 Amount of bone tissue in proximity of the implant after eight weeks in rabbit femur for titanium alloy, NaOH-treated titanium alloy and bioactive glass coated titanium. 1: 250 µm; 2: 1 mm. (Drawn from data obtained from Ref. [14].)

11.7 ANTIMICROBIAL PROPERTIES OF BIOACTIVE GLASSES

The oral microflora consists of several hundred microbial species. Colonisation of certain microorganisms causes caries and periodontitis. Microorganism-associated infection around a dental implant might lead to bone resorption and implant failure. The first antibacterial effects of bioactive glasses against microbes were reported by Stoor *et al.* [15]. Fine-grained S53P4 glass with an average particle size around 20 μm was found to exhibit antibacterial properties when used as a paste with a high particle concentration, 1.7 g of glass per millilitre of bacterial culture solution. The antibacterial effect was attributed to the high pH of the interfacial solution caused by the dissolution of sodium and calcium ions from the glass. The rapid increase of the osmotic pressure in the paste due to ion dissolution was assumed to affect the viability of the bacteria as well. Bioactive S53P4 glass has been proven effective in killing pathogens connected with enamel caries (*Streptococcus mutans*), root caries (*Actinomyces naeslundii*, *S. mutans*) and periodontitis (*Actinobacillus actinomycetemcomitas*, *Porphyromonas gingivalis*, *Prevotella intermedia*).

Silver ions are known to be antimicrobial and they can be introduced into a glass easily (e.g. by substituting Na for Ag). The Ag ions are then released during dissolution, giving the possibility of manufacturing bioactive glasses with bacteria-killing capability. The first silver-containing antibacterial glass was a sol-gel derived composition (wt%): $76SiO_2-19CaO-2P_2O_5-3Ag_2O$ [16]. The minimum concentrations of glass required to kill *Escherichia coli*, *Pseudomonas aeruginosa* and *Staphylococcus aureus* were 1, 0.5 and 0.5 mg/ml, respectively. These three bacteria are associated with biomaterial-related infection sites. Too rapid and high silver dissolution might be cytotoxic to the tissues but, importantly, at these levels, other cells such as human bone cells were not affected. In the same studies, the 45S5 glass did not show any antimicrobial effects. In later studies, however, nanosized 45S5 has been shown to kill *Enterococcus faecalis*, a microorganism associated with failed root canal treatments [17]. Doping surfaces of bioactive glass particles with silver ions by molten ion exchange can be used to tailor the release of the silver ions [18].

Bioactive glasses have been found to inhibit the growth of 29 aerobic and 17 anaerobic clinically important bacteria *in vitro* [19]. The aerobic bacteria were cultured with six melt-quenched and two sol-gel derived glasses, while the anaerobic bacteria were cultured with two melt-quenched and one sol-gel derived glasses. All glasses showed a growth-inhibiting effect, but the bacteria culture time and concentration of glass needed varied with glass composition and bacterial type. The antibacterial effect correlated with both the pH of the solution and the concentration of the alkali ions dissolved from the glasses. In both series, S53P4 glass was found to be the most effective in killing the bacteria. The mechanism of the antibacterial effect of bioactive glasses is not properly understood. However, the observations clearly indicate the potential of bioactive glasses in preventing and treating bacteria-induced infections.

11.8 BIOACTIVE GLASSES IN POLYMER COMPOSITES

Biostable synthetic polymers are used in restorative dentistry. The adhesion and bonding between the dental restoration and the biological tissue are crucial for satisfactory performance. Mixing with glass particles to produce polymer composites offers a feasible method to enhance the osteointegration of the implant. Chapter 9 discussed composites in detail.

The use of fibre-reinforced composites (FRCs) in dental prosthetic devices is steadily increasing. One novel application studied was the replacement of the currently used cylindrical screw-like pure titanium or titanium alloy oral implants with FRC implants. The mismatch of the elastic properties between the metal and the bone might lead to failure of the implant. The use of FRC implants has been motivated by the possibility to tailor the elastic modulus to values close to that of the bone. The FRC implants, fabricated of light-curing E-glass-fibre dimethacrylate composite, were coated with S53P4 glass particles to improve the osteointegration. According to push-out testing from dental plaster, the threaded FRC cones coated with glass particles successfully withstood static load values comparable to human bite forces without fracture [20]. The push-out force was higher for the threaded FRC implants with bioactive glass coating than for the titanium implants. The method for the manufacture of the implant was found to affect the tissue response *in vivo*.

The utilisation of bioresorbable polymers in dental and orthopaedic applications has arisen from the biological ability of the bone tissue to self-repair. Biodegradable polymers can provide temporary mechanical support to the healing tissue and ideally the polymer dissolves at the same rate as the new tissue is regenerated. The most important synthetic bioabsorbable polymers studied for clinical uses are polyglycolide (PGA), polylactides (PLAs), poly(lactic-*co*-glycolide) (PLGA), polycaprolactone

(PCL) and their copolymers [21]. In the composites, glasses are often added as particles or fibres.

In the treatment of bone defects around dental implants, easy application of the filler material is desired. Glass particles are difficult to handle and maintain in the defect site. A composite blended of particles of S53P4 glass and the thermoplastic biodegradable polymer poly(ε-caprolactone-co-DL-lactide) was discovered to be easy to inject into bone defects. It stayed mouldable for a few minutes at body temperature after being heated to $50\,°C$, which is the melting point of the polymer [22]. The composite showed improved bone response *in vivo* when compared to polymer without glass particles.

11.9 BIOACTIVE GLASSES IN GLASS IONOMER CEMENTS

Glass ionomer cements are powdered fluoroaluminosilicate glasses mixed with water-soluble polymers, usually poly(acrylic acid). After mixing, the cement will harden. The glass ionomer cements are adhesive, long-term fluoride-releasing materials used in dentistry as liners and restorations. Their fluoride release improves remineralisation and reduces the solubility of dentine and enamel. The glass ionomer cements bond chemically to enamel and dentine. However, the poor mechanical properties of glass ionomer cements have restricted their use as filling materials mainly to cavities in primary teeth or areas not subjected to high loading. Their low shrinkage has great benefits when they are used to fill cavities.

The first *in vivo* results of the ability of bioactive glass-containing glass ionomer cements to remineralise damaged dentine were reported by Yli-Urpo et al. [23]. The mixture of two commercial glass ionomer cements with powdered bioactive S53P4 glass (diameter $< 45\,\mu m$) was filled into cavities drilled in the teeth of dogs. The results suggested that glass ionomer cements have the potential to mineralise dentine *in vivo* [24]. However, the compressive strength and hardness of glass ionomer cements *in vitro* decreased with increasing amount of bioactive glass. Accordingly, bioactive glass-containing glass ionomer cements were suggested only for applications where the bioactivity can be beneficial, such as root surface fillings and liners, and where high compressive strength is not needed. The decrease in the mechanical properties was attributed mainly to the poor bonding between the polymer and the glass. Modifications in the structure of the polymer have been reported to give better bonding and mechanical properties to the glass ionomer cement containing bioactive glass [25].

11.10 SUMMARY

Bioactive glasses are characterised by their ability to bond chemically to both hard and soft tissues. Although several studies have been conducted in the field, the clinical applications of bioactive glasses in dentistry are limited. Most studies of bioactive glasses deal with bone research. Currently, bioactive glasses find clinical uses in granule form as bone fillers in augmentation of the alveolar ridge.

One of the main challenges for an increased utilisation of bioactive glasses in dentistry is to design products and packages that can be handled and used easily by dental healthcare professionals. Using bioactive glasses as components of thermosetting or light-curing polymers offers a possibility to develop products that can be formed *in situ* to bone fillers, implants or other reconstruction materials. The bioactive glasses aid osteointegration of biostable polymer implants and enhance the regeneration capability of the bone tissue of biodegradable polymer scaffolds.

Fine particulates of bioactive glasses also have potential in toothpastes for treating hypersensitivity, and for enhancing enamel and dentine mineralisation. The ability of small particles of bioactive glasses to reduce the viability of oral microorganisms offers interesting possibilities to develop biomaterials that can be used, for example, in temporary or permanent fillings in treatment of caries.

The dissolution rate and the bioactivity of the glasses can be adjusted with changes in the composition. This offers a possibility to develop novel products for dental applications. For example, slowly dissolving and osteoconductive glass particles and fibres could be of interest for fibre-reinforced composites or in composites with biodegradable polymers.

REFERENCES

[1] Hench, L.L. (1991) Bioceramics: from concept to clinic. *Journal of the American Ceramic Society*, 74, 1487–1510.
[2] Andersson, Ö.H., Karlsson, K.H., Kangasniemi, K. and Yli-Urpo, A. (1988) Model for physical properties and bioactivity of phosphate opal glasses. *Glastechnische Berichte*, 61, 300–305.
[3] Hupa, L., Karlsson, K.H., Aro, H. and Hupa, M. (2010) Comparison of in vitro and in vivo reactions of bioactive glasses. *Glass Technology: European Journal of Glass Science and Technology*, 51, 89–92.

[4] Brink, M., Turunen, T., Happonen, R.P. and Yli-Urpo, A. (1997) Compositional dependence of bioactivity of glasses in the system $Na_2O-K_2O-MgO-CaO-B_2O_3-P_2O_5-SiO_2$. *Journal of Biomedical Materials Research*, 37, 114–121.

[5] Vedel, E., Arstila, H., Ylänen, L. and Hupa, M. (2008) Predicting physical and chemical properties of bioactive glasses from chemical composition. Part I. Viscosity characteristics. *Glass Technology: European Journal of Glass Science and Technology*, 49, 251–259.

[6] Zhang, D., Vedel, E., Hupa, L. and Aro, H.T. (2009) Predicting physical and chemical properties of bioactive glasses from chemical composition. Part III. In vitro reactivity of glasses. *Glass Technology: European Journal of Glass Science and Technology*, 50, 1–8.

[7] Stanley, H.R., Hall, M.B., Clark, A.E. *et al.* (1997) Using 45S5 Bioglass cones as endosseous ridge maintenance implants to prevent alveolar ridge resorption: a 5-year evaluation. *International Journal of Oral and Maxillofacial Implants*, 12, 95–105.

[8] Turunen, T., Peltola, J., Yli-Urpo, A. and Happonen, R.P. (2004) Bioactive glass granules as a bone adjunctive material in maxillary sinus floor augmentation. *Clinical Oral Implants Research*, 15, 135–141.

[9] Forsback, A.P., Areva, S. and Salonen, J.I. (2004) Mineralization of dentin induced by treatment with bioactive glass S53P4 in vitro. *Acta Odontologica Scandinavica*, 62, 14–20.

[10] Earl, J.S., Leary, R.K., Muller, K.H. *et al.* (2011) Physical and chemical characterization of dentin surface, following treatment with NovaMin® technology. *Journal of Clinical Dentistry*, 22, 2–67.

[11] Mneimne, M., Hill, R.G., Bushby, A.J. and Brauer, D.S. (2011) High phosphate content significantly increases apatite formation of fluoride-containing bioactive glasses. *Acta Biomaterialia*, 7, 1827–1834.

[12] Dong, Z., Chang, J., Zhou, Y. and Lin, K. (2011) In vitro mineralization of human dental enamel by bioactive glass. *Journal of Materials Science*, 46, 1591–1596.

[13] Gjorgievska, E. and Nicholson, J.W. (2011) Prevention of enamel demineralization after tooth bleaching by bioactive glass incorporated into toothpaste. *Australian Dental Journal*, 56, 193–200.

[14] Moritz, N., Rossi, S., Vedel, E. *et al.* (2004) Implants coated with bioactive glass by CO_2-laser, an in vivo study. *Journal of Materials Science: Materials in Medicine*, 15, 795–804.

[15] Stoor, P., Söderling, E., Yli-Urpo, A. and Salonen, J. (1998) Antibacterial effects of a bioactive glass paste on oral microorganisms. *Acta Odontologica Scandinavica*, 48, 161–165.

[16] Bellantone, M., Colman, N.J. and Hench, L.L. (2000) Bacteriostatic action of a novel four-component bioactive glass. *Journal of Biomedical Materials Research*, 51, 484–490.

[17] Waltimo, T., Brunner, T.J., Vollenweider, M. *et al.* (2007) Antimicrobial effect of nanometric bioactive glass 45S5. *Journal of Dental Research*, 86, 754–757.

[18] Verné, E., Miola, M., Vitale-Brovarone, C. *et al.* (2009) Surface silver-doping of biocompatible glass to induce antibacterial properties. Part I: massive glass. *Journal of Materials Science: Materials in Medicine*, 20, 733–740.

[19] Zhang, D., Leppäranta, O., Munukka, E. et al. (2010) Antibacterial effects and dissolutio behavior of six bioactive glasses. Journal of Biomedical Materials Research, Part A, 93A, 475–483.

[20] Munukka, E., Leppäranta, O., Korkeamäki, M. et al. (2008) Bactericidal effects of bioactive glasses on clinically important aerobic bacteria. Journal of Materials Science: Materials in Medicine, 19, 27–32.

[21] Ballo, A.M., Kokkari, A.K., Meretoja, V.V. et al. (2007) Load bearing capacity of bone anchored fiber-reinforced composite device. Journal of Materials Science: Materials in Medicine, 18, 2025–2031.

[22] Rezwan, K., Chen, Q.Z., Baker, J.J. and Boccaccini, A.R. (2006) Biodegradable and bioactive porous polymer/inorganic composite scaffolds for bone tissue engineering. Biomaterials, 27, 3413–3431.

[23] Närhi, T.O., Jansen, J.A., Jaakkola, T. et al. (2003) Bone response to degradable thermoplastic composite in rabbits. Biomaterials, 24, 1697–1704.

[24] Yli-Urpo, H., Närhi, M. and Närhi, T. (2005) Compound changes and tooth mineralization effects of glass ionomer cements containing bioactive glass (S53P4), an in vivo study. Biomaterials, 26, 5934–5941.

[25] Yli-Urpo, H., Lassila, L.V.J., Närhi, T. and Vallittu, P.K. (2005) Compressive strength and surface characterization of glass ionomer cements modified by particles of bioactive glasses. Dental Materials, 21, 201–209.

[26] Xie, D., Zhao, J., Weng, Y. et al. (2008) Bioactive glass-ionomer cement with potential therapeutic function to dentin capping mineralization. European Journal of Oral Sciences, 116, 479–487.

12

Bioactive Glass as Synthetic Bone Grafts and Scaffolds for Tissue Engineering

Julian R. Jones

Department of Materials, Imperial College London, London, UK

12.1 INTRODUCTION

Bioactive glass was invented in 1969 by Larry Hench. Since then, bioactive glasses have been seen to bond to bone and to degrade safely, releasing bioactive ions that can stimulate stem cells and bone cells to produce new bone. However, few products have been released to market (Chapter 2). Bioactive glass has been lagging behind other bioceramics in terms of numbers of products and amount of use worldwide. What are the reasons for this? It is not down to performance: *in vivo* studies show that bioactive glasses outperform bioactive ceramics such as calcium phosphates and synthetic hydroxyapatites (Chapter 2). So what has been the problem?

Bio-Glasses: An Introduction, First Edition. Edited by Julian R. Jones and Alexis G. Clare.
© 2012 John Wiley & Sons, Ltd. Published 2012 by John Wiley & Sons, Ltd.

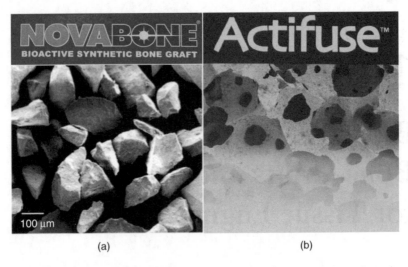

Figure 12.1 Scanning electron microscope images of commercial synthetic bone grafts: (a) NovaBone and (b) Actifuse.

Bioactive glass began commercial life as a particulate bone filler in dental applications in 1995 and then became an orthopaedic product a few years later. Surgeons mix the sachet of bioactive glass powder with blood from the bone defect of the patient and push the mixture into the defect as a putty. One of the main problems for lack of widespread use of bioactive glass in orthopaedic applications has been the lack of products available. Several compositions, such as Bioglass® – the original Hench formulation – available as PerioGlas® and NovaBone® (NovaBone Products, USA), Biogran® (Orthovita, USA), BonAlive® (BonAlive, Finland) and StronBone™ (RepRegen, UK) exist, but they are all particulates (Figure 12.1a). So let us take a look at the current market-leading synthetic bone graft: Actifuse® (Apatech, UK). Actifuse is a synthetic hydroxyapatite. What sets it apart from other synthetic hydroxapatites (of which there are many commercial products) is that it contains a small amount (0.8 wt%) of silicon. It is silicon-substituted hydroxyapatite. Traditional synthetic apatites are highly crystalline ceramics that degrade slowly and bond to bone slowly. The small amount of silicon in the Actifuse formulation creates defects, increasing the number of grain boundaries and reducing crystallinity just enough to increase degradation. Silicon was chosen because of studies that showed that a diet containing silicon is necessary for healthy bone growth and because bioactive glass was known to be more bioactive than the bioactive ceramics. One reason for the latter was that the glasses release

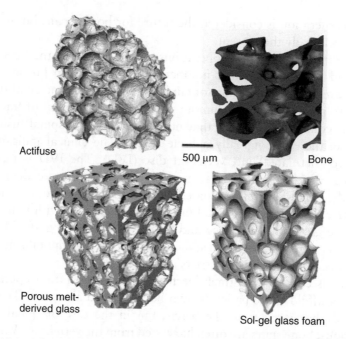

Figure 12.2   Three-dimensional images, obtained using X-ray microtomography, of porous networks in Actifuse, cancellous bone and bioactive glass foams (melt-derived and sol-gel derived). (Images by Sheng Yue. Copyright (2012) Sheng Yue.)

soluble silica, which stimulates cells to produce more bone. What sets Actifuse apart from bioactive glass is that it consists of porous granules (Figure 12.1b). The pores are of a similar size to those in porous bone (Figure 12.2), which means the granules mimic the macrostructure of the cancellous or trabecular bone.

12.2 SYNTHETIC BONE GRAFTS AND REGENERATIVE MEDICINE

Regenerative medicine is using materials (implants), cells, drugs or a combination of these to stimulate the body to regenerate diseased or damaged tissue to its original state and function. For example, a tumour is removed from a bone, leaving a large defect. The bone would have been able to heal itself if the defect was small, like a routine fracture, but if the gap is large, the cells cannot 'sense' the other side of the defect and the defect remains, or is filled with soft tissue. A temporary template (scaffold) is needed that can help the cells. The scaffold acts as a framework and guide for the cells that usually regenerate the bone.

Bone regeneration is considered the future for bone repair, but in some ways it is already the present.

Half a million bone graft operations are performed in the USA every year, and just over half that number are carried out in Europe. Bone defects are caused by a range of clinical indications: congenital defects (e.g. cleft palates), trauma, tumour removal or non-union of fractures. Another common procedure that needs bone graft is spinal fusion. In this case, there is not really a bone defect. The clinical indication is a slipped (herniated) intervertebral disc (IVD). The IVD is made of cartilage, which regenerates at a much, much lower rate than bone. When the IVD is severely herniated and has to be removed, treatment is to replace the IVD with a metal or polymer cage filled with bone graft. The bone grows into the cage and fuses with the bone graft. The idea is that the implant fuses the neighbouring vertebrae, immobilising them. This reduces pain but also restricts movement.

Surgeons currently routinely perform bone regeneration operations, but the scaffold is the patient's own bone, termed autograft. The bone is usually harvested from the pelvis, but in the case of spinal fusion operations, bone spurs are often harvested from the vertebrae. Autograft bone is an ideal material for regenerating bone.

Unfortunately, there are many drawbacks to using the patient's bone. The most important is that there is limited supply of bone that can be utilised without causing problems at the host site. The body is not wasteful in bone production, so there is not much excess bone available. One of the main functions of bone is to support load. If a situation occurs where excess bone is made, the excess bone is likely not to be under load. If this happens, the body will take the bone away by osteoclast action. This is why astronauts need to exercise excessively in space to prevent loss of bone density, as the lack of gravity means bones are not loaded. This is an important consideration for bone graft design.

Harvesting healthy bone for grafting also creates another defect that needs to be healed (without the help of more bone). The healing of this defect is extremely painful and long. Patients generally feel a lot of pain at the donor site and one in four patients will experience complications at the defect site long after the operation. Some will require revision surgery of the harvest site, which is not ideal for the patient, the healthcare service or the economy.

These problems are the driving force for the need for synthetic bone grafts that can perform as well as (or better than) autograft. The devices must regenerate bone defects without the need for graft operations so that patients can heal quickly and pain-free, returning to their normal lives and, importantly for the economy, place of work, more quickly.

An artificial bone graft that can replace the need for grafts would have a massive impact on the global economy. The device market itself is thought to be worth $2 billion per year without taking into account the economic impact of reduced operating costs and faster recovery times.

Many artificial bone grafts are designed to replace or augment the bone and stay there for a long time rather than regenerate the bone to its original state and function. Examples are porous titanium or tantalum metal constructs. Metallic scaffolds have the advantage of high strength and toughness. Toughness (resistance to crack propagation) is important when the scaffold is to be exposed to cyclic loads, which are commonplace in skeletal tissue. However, metals are usually bio-inert and intrinsically not bioactive, so fibrous encapsulation may occur. The metal will also stay in place long-term, meaning that naturally healthy bone will never re-form. The body is likely eventually to reject the implant, but time scales are variable.

12.3 DESIGN CRITERIA FOR AN IDEAL SYNTHETIC BONE GRAFT

Surgeons ultimately want something that works and can be implanted with ease. They would like to be able to take a packet off the shelf, remove the implant, shape it, press or inject it into the defect, and watch as it fills the space. They would like the implant to take load and share load with the host bone, immediately. This is so the patient can load their bones, keeping them healthy, but also so bed space is not occupied for too long. Then, over time, they would like the scaffold to disappear as the bone regrows.

An ideal synthetic bone graft would regenerate a bone defect and leave no trace of an implant, and it must possess the following characteristics:

(a) be biocompatible and bioactive, promoting bone formation, and bond to the bone without soft tissue encapsulation;
(b) act as a template for bone growth, with an interconnected porous structure that allows cell migration and vascularisation;
(c) biodegrade safely in the body and have a controllable degradation rate;
(d) have mechanical properties similar to those of the host bone;

(e) be produced by a process that allows the scaffold to be shaped to fit a range of defect geometries and be up-scalable for mass production; and

(f) be sterilisable and meet the regulatory requirements for clinical use.

There are numerous synthetic bone graft materials as commercial products, but none that fulfil all the criteria listed above. Metals are not bioactive and generally do not degrade, although magnesium is being investigated as a potentially biodegradable metal for stent applications. Bioceramics (calcium sulfates and calcium phosphates) are usually chosen because of their long track record in clinical use. Bioactive glasses have the potential to fulfil most of these criteria. But currently their commercial products are limited to particles. Why is this?

12.4 BIOGLASS AND THE COMPLICATION OF CRYSTALLISATION DURING SINTERING

All processing techniques for making porous scaffolds from particles involve sintering. Sintering is the fusion of particles at high temperature (Figure 12.3). Raising the temperature above the glass transition temperature (T_g) causes local flow of the glass and allows particles to fuse. However, to maintain the amorphous glass structure and properties, the temperature must not be raised above the crystallisation onset temperature $(T_{c,onset})$. Unfortunately for the original Bioglass composition (46.1 mol% SiO_2, 24.4 mol% Na_2O, 26.9 mol% CaO and 2.6 mol% P_2O_5), the low silica content causes T_g and $T_{c,onset}$ to be too close together, so it is not possible to sinter the glass without crystallising, which leads to the formation of a glass-ceramic and reduction in bioactivity. Therefore, only Bioglass particles are available commercially (Figure 12.1). Only now have researchers been able to understand the relationship between glass structure, composition, glass

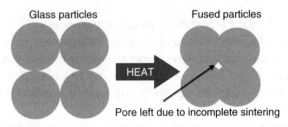

Glass particles   Fused particles

HEAT

Pore left due to incomplete sintering

Figure 12.3   Schematic of the principles of sintering glass particles.

transition, crystallisation and bioactivity. In other words, it is quite a challenge to design a glass composition that can be sintered without crystallising but also remains bioactive. One line of thought is that the network connectivity should be approximately 2 (like Bioglass) for a glass to be bioactive (Chapter 2). However, glasses with network connectivity of 2 have the problem of easy crystallisation. Increasing the silica content reduces the tendency of a glass to crystallise, but this reduces the degradation rate and bioactivity. We therefore need to 'trick' the glass network by incorporating a variety of network modifiers, substituting for calcium and sodium, to keep network connectivity constant. The variety of modifiers makes crystallisation energetically unfavourable, as the structure is more difficult to organise.

New compositions have been designed not to crystallise on sintering. One is 13-93 (54.6 mol% SiO_2, 6 mol% Na_2O, 22.1 mol% CaO, 1.7 mol% P_2O_5, 7.9 mol% K_2O, 7.7 mol% MgO), which was developed in Finland by Brink and co-workers. This glass takes seven days to form a hydroxycarbonate apatite layer in simulated body fluid tests (Bioglass particles formed the same layer within 8 hours). This is because the network connectivity is higher in glass composition 13-93 compared to 45S5 Bioglass owing to the increased silica content.

In order to obtain a similar result without compromising bioactivity, ICIE16 (49.46 mol% SiO_2, 36.27 mol% CaO, 6.6 mol% Na_2O, 1.07 mol% P_2O_5 and 6.6 mol% K_2O) was developed by Elgayar and co-workers.

12.5 MAKING POROUS GLASSES

Sintering alone cannot make pores large enough to create a pore network that can encourage vascularised bone growth. Pores that are left behind (Figure 12.3) are considered defects and sources of weakness (crack nucleation sites and sites of stress concentration). The aim is to create large pores with diameters in excess of 500 μm with interconnects greater than 100 μm, while having highly sintered struts that provide as much strength as possible. The aim is to mimic the porous structure of cancellous bone (Figure 12.2).

12.5.1 Space Holder Method

The most common method for making porous ceramics is to take a particulate and pack the particles around a sacrificial template of some

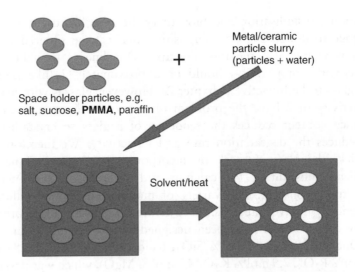

Figure 12.4   Schematic of the porogen or space holder method for producing porous materials.

kind (Figure 12.4). During sintering, the particles will fuse together. The template can be either washed out or burned out, depending on what is used, leaving pores. The pore size and interconnectivity depend on the template.

The space holder or porogen method is the most common. Sacrificial particles are used, either soluble particles, for example, salt or sucrose, or combustible, for example, poly(methyl methacrylate) (PMMA) microbeads (Figure 12.4). Combustible polymers are usually used for ceramic synthesis. However, the burning out of the sacrificial polymer step must be done carefully. For processing bioactive glass, washing with aqueous solutions is avoided to prevent the start of glass dissolution and the bioactivity mechanism cascade. However, if not enough oxygen reaches the combustible polymer, residual carbon will be left behind, leaving a black colour and reducing the sintering efficiency. This is known as coring. The use of PMMA reduces coring because it leaves little residue as it burns. High oxygen content in the furnace and sintering samples as small as possible (reducing the path length that oxygen molecules would need to travel) can also help to reduce coring. The advantage of the space holder technique is that it is simple and can be up-scaled to production easily. Pore size is largely determined by particle size of the sacrificial polymer, but it is difficult to maintain a homogeneous distribution of the polymer spheres and therefore interconnectivity is low and poorly controlled (Figure 12.5).

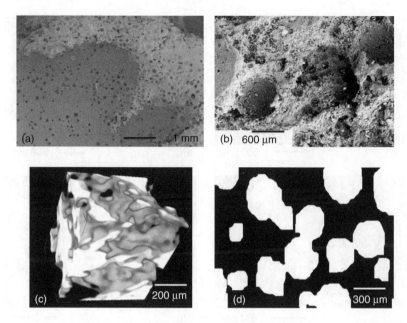

Figure 12.5 Images of porous bioactive glasses produced by the space holder method, using PMMA microspheres (diameter $\sim$500 μm) at a glass-to-polymer ratio of 50 : 50. (a) Low-magnification SEM image showing isolated spherical pores. (b) SEM image showing isolated pores (not interconnected). (c) 3D reconstruction from X-ray microtomography (μCT) imaging, showing irregularity of pores. (d) 2D μCT projection showing isolated pores. (Images by Zoe Wu and Sheng Yue. Copyright (2012) Zoe Wu and Sheng Yue.)

12.5.2 Polymer Foam Replication

Interconnectivity is improved through using sacrificial polyurethane foams rather than spheres. Polyurethane foams are used to make common open-pore (open-cell) foams such as those used in sofas and armchairs. They are readily available in different ranges of pore sizes, although they are usually quoted in pores per inch. Figure 12.6(a) shows a scanning electron microscope (SEM) image of a polyurethane foam. The struts are thin and the pores are large and well connected, so much so that it is difficult to tell what is an interconnection and what is a pore. The polymer foam can be immersed in slurries of glass or ceramic particles so that the particles coat the polymer foam. The aim is that, after sintering, the glass will take the shape of the foam. The main challenge in the process is to ensure that the polymer is well coated but not full of excess particles. If there are excess particles, they will block the pores. The common way to remove excess powder is to squeeze it

Figure 12.6 SEM images relating to the polymer foam replication process: (a) a sacrificial polyurethane foam template; (b) a porous glass foam after removal of the polymer template and sintering; (c) a porous glass of lower pore size; and (d) cross-section of a hollow strut caused by the polymer removal. (Images by Zoe Wu. Copyright (2012) Zoe Wu.)

out of the foam, literally between two fingers. Industrial companies must have an automated process for this as reticulated ceramic foams are mass produced for other applications. After the excess powder is removed, the foams are heated at $250\,^\circ$C to burn out the organic components (pyrolysis) and sintered for 3 hours. For glasses, the sintering temperature is chosen depending on the glass composition, as it must be in the sintering window between the T_g and T_c (usually around $700\,^\circ$C). Figure 12.6 shows SEM images of the polymer foam (Figure 12.6a) and the resulting glass scaffold (Figure 12.6b). Polymer foam replication is successful in that it produces a very open interconnected structure. Polymer foams are easy to produce or purchase, and pore size is very much determined by the polymer foam specifications, which means foams of different pore sizes can be easily produced. Figure 12.6(c) shows a glass foam that was produced from a polymer foam of lower pore size. However, up-scaling is challenging. Polymer removal also leaves hollow foam struts (Figure 12.6d), which means that mechanical properties are low.

12.5.3 Direct Foaming

A way to avoid having hollow struts is to directly foam the slurry so a polymer foam is not needed. This technique involves the use of surfactants to stabilise bubbles that are created in a liquid by vigorous agitation. The bubbles must then be gelled (solidified) to maintain the porous structure. This is a key step in the process, as the bubbles must be maintained, but solidified. The process is similar to what is used to produce Actifuse, and is the latest technique for producing porous bioactive glasses with similar interconnected pore structures and mechanical strengths to porous bone.

For direct foaming, either melt-derived or sol-gel glasses can be used. Melt-derived glasses are foamed by the gel-cast foaming process and sol-gel by the sol-gel foaming processes. The processes have many similarities. The main ones are that in both cases a solution or slurry is foamed under vigorous agitation with a surfactant to form bubbles. The bubbles are gelled and are poured into moulds immediately prior to gelling. The main differences are that the melt-derived glass gel-cast foaming process uses *in situ* polymerisation to gel the bubbles. The sol-gel process is different in that, rather than needing a polymer to do the gelling, it gels itself, which simplifies the process. Surfactants are 'surface active agents'. They are molecules that have a hydrophilic end and a hydrophobic end and are the active ingredients of detergents. When surfactants are added to water, they lower the surface tension because the hydrophilic end of the molecule affiliates itself with the water, and the hydrophobic end is in the air. This stabilises the bubbles (Figure 12.7) that grow by agitation.

12.5.4 Gel-Cast Foaming

For melt-derived glass, fine particles ($<38\ \mu m$) of a sinterable composition, such as 13-93 or ICIE16, are added to water to produce a slurry. A surfactant is then added and the slurry is foamed under vigorous agitation. This could be with a whisk, rather like making a meringue. Figure 12.8 shows a schematic of the process. For the process to succeed, the viscosity must increase and then the slurry must be gelled to bind the particles around the bubbles and permanently fix them in place. In the gel-cast foaming process, the gelation is achieved by *in situ* polymerisation, that is, a polymer is formed while the foaming is being carried out. Monomers (usually acrylates) are usually used, which are polymerised by

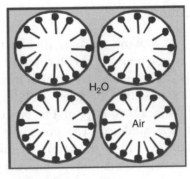

- ● Surfactant molecule
- ● Hydrophilic polar end
- — Non-polar end

Figure 12.7 Schematic on the role of surfactant in stabilising air bubbles in agitated water.

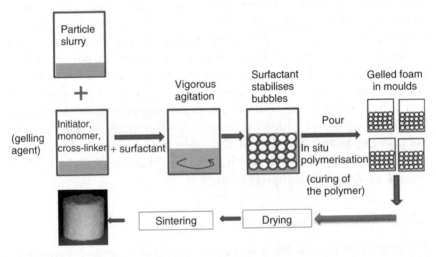

Figure 12.8 Schematic of the gel-cast foaming process for bioactive glass scaffold production.

mixing with an initiator and catalyst. As the polymerisation progresses, the viscosity increases until a gel (a solid covalent network containing water) forms. Just prior to gelation, the foam is poured into a mould.

The pouring window is short: too early, and the foam cannot hold its own weight and will collapse; too late, and it will gel in the foaming vessel. The surfactant must be of suitable type and be homogeneously dispersed to obtain spherical pores. To obtain an interconnected pore network, the bubbles must be large and touching each other in the solution until gelation. On gelation, the surfactant films must rupture, opening up interconnecting channels between the bubbles, which now become the pores. After gelation, the foam is a composite of glass particles within the newly formed polymer matrix (Figure 12.9).

In order to make the porous glass, the polymer has to be removed. Polymer removal and sintering occur in the same heat treatment procedure. The composite is usually held at around 300 °C to remove the polymer. At this point, the particles are effectively balancing on each other in the shape of a foam. As the temperature increases above T_g, the particles begin to sinter together. The sintering temperature depends on the sintering window of the glass composition being used, but is usually around 700 °C. A 3D image of a gel-cast foam scaffold can be seen in Figure 12.2. The scaffolds have large interconnecting pores without hollow struts. Figure 12.10 shows SEM images of a bioactive glass scaffold after sintering. Note the smoothness of the foam surface at higher magnification after sintering, which shows that sintering has run to completion. The amount of glass loading in the slurry is a critical

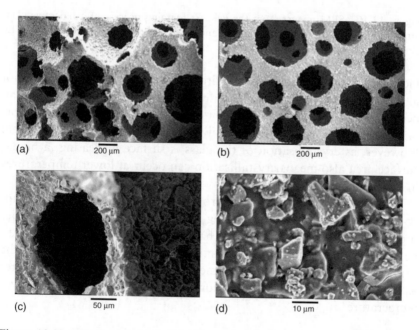

(a)   200 μm   (b)   200 μm

(c)   50 μm   (d)   10 μm

Figure 12.9 SEM images from the gel-cast foaming process of a bioactive glass after foaming and gelation of glass particles dispersed in a polymer foam: (a, b) low magnification; and (c, d) higher magnification, showing individual particles in the polymer matrix. (Images by Zoe Wu. Copyright (2012) Zoe Wu.)

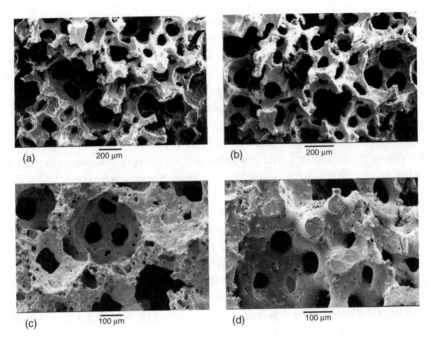

Figure 12.10 SEM images from the gel-cast foaming process of a bioactive glass after sintering: (a, b) low magnification; and (c, d) higher magnification. (Images by Zoe Wu. Copyright (2012) Zoe Wu.)

factor in getting a good foam: too little glass means the particles are not in contact with each other, and the foam will slump before sintering can occur; and too much glass is difficult to foam. Particle size is also important: small particles sinter more easily as they have a higher surface area. However, as crystallisation of the glass is surface nucleating, a higher surface area also means crystallisation can occur at lower temperatures in smaller particles.

The gel-cast foaming process produces excellent scaffolds, but upscaling for production is challenging. Another challenge is that, in order for a surfactant to function, water has to be present, so the glass has to be in a slurry of water. This means the glass will start to react with the water. Although the amount of time the glass is exposed to the water is short, it can trigger crystallisation of the glass to occur at a lower temperature.

12.5.5 Sol-Gel Foaming Process

The sol-gel process involves the hydrolysis of alkoxide precursors to create a sol (Chapter 3). The sol can be considered as a solution of silica species that undergo polycondensation to form silica nanoparticles, which then coalesce; further condensation links them together during gelation (under acidic catalysis). Calcium is usually incorporated using calcium nitrate. The gels are then dried and heated to at least 600 °C to remove the nitrates from the calcium nitrate. During the thermal processing, the coalesced nanoparticles sinter together, leaving interstitial nanoporosity. The nanopores are usually in the range of 1–20 nm diameter and can be tailored during processing by controlling the pH of the catalyst, the nominal composition and the final temperature. It is, however, difficult to produce large crack-free monoliths (greater than 10 mm thickness) because driving off the water, organics and nitrates causes capillary stresses that result in cracking.

Larger pieces can be made using the sol-gel foaming process, as the pores that are introduced reduce the distance that water molecules need to travel. To produce a porous glass, the sol is foamed under vigorous agitation (Figure 12.11). The viscosity of the sol increases until it becomes a solid gel, and therefore no polymeric gelation agent is needed. However, an extra catalyst is needed. Gelation normally

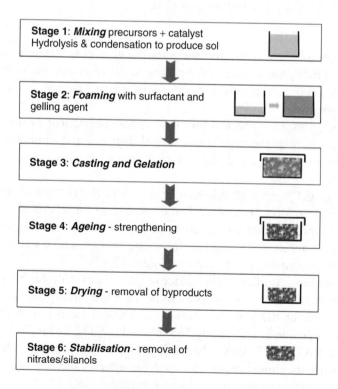

Stage 1: *Mixing* precursors + catalyst
Hydrolysis & condensation to produce sol

Stage 2: *Foaming* with surfactant and gelling agent

Stage 3: *Casting and Gelation*

Stage 4: *Ageing* - strengthening

Stage 5: *Drying* - removal of byproducts

Stage 6: *Stabilisation* - removal of nitrates/silanols

Figure 12.11 Schematic of the sol-gel foaming process.

Figure 12.12   SEM image of the nanotopography of a 70S30C (70 mol% SiO$_2$, 30 mol% CaO) sol-gel derived bioactive glass.

takes around three days, and for a foaming process, a few minutes is needed. Hydrofluoric acid is the gelling agent of choice because, when gelation occurs, it occurs rapidly, allowing the porous foam to gel homogeneously.

On gelation, the spherical bubbles become permanent in the gel. As drainage occurs in the foam struts, the gel shrinks and the bubbles merge, interconnects open up at the point of contact between neighbouring bubbles.

The sol-gel foam scaffolds have a hierarchical structure of intercon-nected macropores (Figure 12.2 and see Figure 14 in colour section), which mimic the porous structure of cancellous bone and allow the scaffold to act as a 3D template for tissue growth, and a nanoporosity that allows control of degradation (Figure 12.12).

Cell response studies on the bioactive glass foam scaffolds have found that primary human osteoblasts lay down mineralised immature bone tissue, without the need for additional growth factors or hormones. Glass compositions are usually 58S (60 mol% SiO$_2$, 26 mol% CaO and 4 mol% P$_2$O$_5$) or 70S30C (70 mol% SiO$_2$, 30 mol% CaO). However, other network modifiers can be used for added functionality, such as strontium (anti-osteoporosis) or silver (antibacterial).

12.5.6 Solid Freeform Fabrication

Solid freeform fabrication techniques are a collection of techniques that can build objects in almost any net shape, by depositing material layer by layer. The method is often also known as rapid prototyping. The advantage of these techniques over foaming is that the scaffold structure is dictated by a computer that controls the device that lays down the material. This means that scaffolds can theoretically be produced in any design as dictated by a computer-aided design (CAD) file. The CAD file could even be generated from a computer-assisted tomography (CAT) scan of a tissue, allowing complete replication of the structure of a tissue. However, in reality, not all materials can be used directly in solid freeform fabrication. Thanks to new melt-derived bioactive glass compositions, bioactive glass scaffolds are produced by a printing process called robocasting. The scaffolds produced had thick struts (>50 μm) and pores in excess of 500 μm (Figure 12.13 and see Figure 13 in colour section). The alignment of the rows of struts was so accurate that compressive strengths of more than 150 MPa were achieved in the direction of the pore channels (three times that perpendicular to the pore channel directions), with 60% porosity. This is similar to the strength of cortical bone. The composition used was 6P53B (51.9 mol% SiO_2, 9.8 mol% Na_2O, 1.8 mol% K_2O, 15.0 mol% MgO, 19.0 mol% CaO, 2.5 mol% P_2O_5), with a particle size of $D_{50} = 1.2$ μm. Inks were created by mixing 30 vol% glass particles in 20 wt% Pluronic F-127 solution.

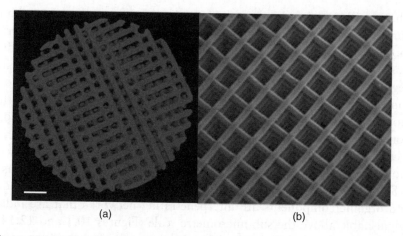

(a)   (b)

Figure 12.13 Bioactive glass scaffolds produced by the robocasting solid freeform fabrication method: (a) X-ray microtomography image and (b) SEM image. (Image provided courtesy of E. Saiz and Q. Fu. Copyright (2012) E. Saiz and Q. Fu.)

Glass scaffolds were fabricated by extruding the inks through a 100 μm syringe nozzle using a robotic deposition device. The viscosity of the ink is critical. The inks were printed on an alumina substrate in a reservoir of non-wetting oil. The scaffolds were air-dried for 24 hours and subjected to a controlled heat treatment to decompose the organics and sinter the glass particles (700 °C).

12.5.7 Summary of Bioactive Glass Scaffold Processing

Bioactive glass scaffolds have been synthesised with foam-like pore networks resembling the structure of cancellous bone. Compressive strengths of 2–15 MPa have been obtained in bioactive glass foam scaffolds while maintaining modal interconnect diameters above 100 μm (>80% porosity). Using solid freeform fabrication, higher compressive strengths were obtained (>150 MPa at 60% porosity), but the architecture was less similar to native bone. Although the compressive strength of glass scaffolds may be suitable for bone grafts in applications where the load is compressive and not cyclic, bioactive glass scaffolds suffer similar problems to other bioceramics: they are brittle. For certain applications, and to replace the need for autografts, improved toughness is needed.

12.6 THE FUTURE: POROUS HYBRIDS

Chapter 10 explains that bioactive glasses must be made more tough if they are to be used in sites that will be under cyclic loading. Chapter 9 showed that there have been several attempts to combine bioactive glasses with biodegradable polymers to create composite scaffolds with degradability, bioactivity and toughness. Chapter 10 discusses how conventional composites are flawed as synthetic bone grafts because the bioactive particles are generally covered by the polymer matrix. The host bone will therefore not come into contact with the glass. This may be rectified as the polymer phase begins to degrade and the glass is exposed. However, the polymer often degrades much more rapidly than the glass.

Sol-gel hybrids are different from composites in that the inorganic and organic components are interpenetrating networks that are indistinguishable above the sub-micrometre scale (Figures 10.1 and 12.14). Because the sol-gel process is initially at room temperature, polymers can be incorporated into the sol so that the polymer network is incorporated as the silica network forms. This nanoscale interaction can produce

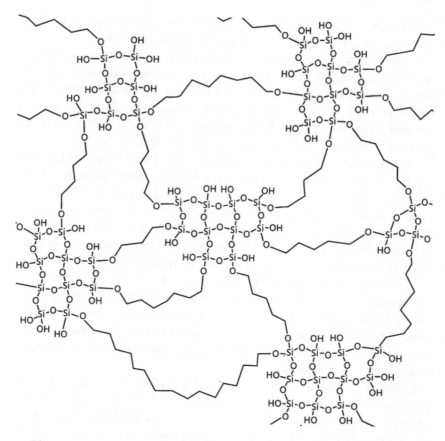

Figure 12.14   Schematic of the structure of a sol-gel silica–organic hybrid.

unique properties. Fine control of the degradation rate and mechanical properties can be achieved when covalent bonds are formed between the organic and inorganic chains (a class II hybrid, Figure 10.6).

Although the final processing temperature of the sol-gel hybrid process is lower (40–90 °C) than that of the sol-gel glass process (>600 °C), the foaming process can still be incorporated. However, there are complex chemistry challenges associated with this procedure, including which polymer to use and how to incorporate calcium.

Traditionally, calcium nitrate has been used as a precursor and dona-tor of calcium into the inorganic network. However, temperatures of at least 600 °C are needed to drive off the nitrate by-products that are toxic to cells. The temperature must also reach 450 °C before cal-cium nitrate dissociates and calcium is incorporated into the network.

Hybrids cannot be heated to high temperatures, as the polymer will be damaged. Therefore, another calcium precursor is needed.

Many bioresorbable polymers, for example, poly(lactides), cannot be simply introduced into the sol, as they are not soluble in the sol. However, they can be functionalised so that not only are they incorporated in the sol–gel process, but also they can form covalent bonds with the silica network, creating a class II hybrid material. The functionalisation of the polymer involves the introduction of coupling agents.

One example is silica–poly(ε-caprolactone) (PCL) hybrids. Hydroxyl groups at either end of poly(ε-caprolactone diol) polymer chains can be reacted with 3-isocyanatopropyl triethoxysilane (IPTS). This process results in a polymer end-capped with a triethoxysilyl group. When the end-capped PCL is introduced into a sol, the siloxane groups hydrolyse and then Si–OH groups from the end-capped polymer condense with the Si–OH groups from the hydrolysed TEOS in the sol to yield an interconnected PCL–silica network.

Alternatives to conventional polyesters are natural polymers, which can be a closer mimic of bone's natural structure. Bone contains collagen, which is a structural protein with a triple helix of polypeptides (amino acid chains), giving it excellent mechanical strength (the structure is analogous to that of rope).

Collagen would therefore be an ideal choice to use as a natural polymer in hybrid synthesis. Unfortunately, the triple helix structure makes it very insoluble. It will dissolve in acetic acid, but only in very low concentrations. Therefore, it is not possible to produce a hybrid with significant amounts of polymer using collagen. Where to source the collagen is also an issue. Currently, it cannot be synthesised in any significant quantity, so it must be sourced from animals. Although collagen is unlikely to be rejected by a patient's body, patients may refuse an implant on religious or moral grounds owing to the animal species from which it originates, for example, bovine (cow) or porcine (pig).

One of the great benefits of natural polypeptides is that they can be functionalised. Gelatin has great potential as it is hydrolysed collagen and it is water-soluble. It also contains –COOH groups (carboxylic acid), –NH and –NH$_2$ groups along its backbone that are available for functionalisation. This time, glycidoxypropyl trimethoxysilane (GPTMS) is used as the coupling agent. The glycidol group (expoxy ring) can open and react with the functional groups on the polymer chain. The functionalised polymer has short molecules bonded to it with Si–OH groups on the end of them, ready to undergo condensation with other Si–OH groups from the silica in the sol, to form Si–O–Si bonds. Any polypeptide

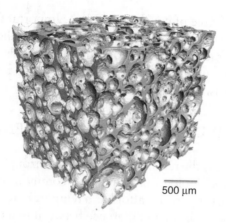

Figure 12.15 Three-dimensional X-ray microtomography image of a sol-gel foam hybrid. (Image by Sheng Yue. Copyright (2012) Sheng Yue.)

can be functionalised and incorporated. An example is poly(γ-glutamic acid) (γ-PGA), which is a much simpler polypeptide than gelatin; γ-PGA is synthesised by a biotechnology route, that is, produced by bacteria.

Another popular natural polymer is chitosan, a polysaccharide derived from crustacean shells, which contains –OH and –NH$_2$ groups.

Class II hybrids of silica–gelatin, silica–γ-PGA and silica–chitosan have been produced with several methods of adding calcium. They have also been foamed to produce porous scaffolds (Figure 12.15). A schematic of the process is shown in Figure 12.16. In some processes, drying is carried out at low temperatures; in others, (preferably) freeze drying quickly removes the water and other by-products of condensation. Scaffolds can be made with stiffnesses ranging from that of a polymer to that of a glass, or anywhere in between, by controlling the percentage of

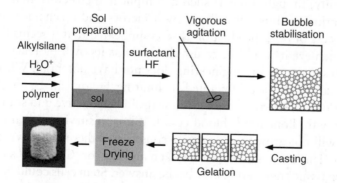

Figure 12.16 A schematic of the sol-gel foaming process for hybrids.

organic and the degree of covalent coupling. Importantly, release of the polymers can be coupled to the degradation of the silica, meaning that the materials degrade as one material, or true hybrids. The future may well be the use of human recombinant proteins, but much development is needed to increase their yields.

12.7 BIOACTIVE GLASSES AND TISSUE ENGINEERING

Tissue engineering is the regeneration of tissues through the combination of engineering and biology principles. A common strategy for bone tissue engineering is to use a scaffold as a temporary template for cells to produce new tissue. The main difference between tissue engineering and a synthetic bone graft that regenerates bone is that tissue engineering is partly done in a laboratory. A scaffold can be seeded with cells in a laboratory and encouraged to grow on the scaffold. The cells must penetrate the scaffold, attach and produce bone matrix. A choice has to be made as to the point at which the construct is implanted: it could be implanted immediately after the cells are seeded; after a specified time; or once an entire tissue has been grown and the scaffold degraded, so only tissue is implanted. Ideal tissue engineering would be growing entire replacement parts (organs) so that they are ready for use, using a patient's own cells to prevent rejection. This has only so far been achieved with skin, using polymer scaffolds, but the skin produced does not contain many of the components of natural skin, for example, sweat glands, pores, pigment cells and hair follicles. A big advantage of growing a large piece of tissue and allowing the scaffold to degrade is that the surgeon will not be implanting any synthetic material, so some of the regulatory issues for medical devices can be avoided.

In reality, though, bone is such a complicated structure that needs a combination of different cells, growth factors and mechanical stimuli. The process of bone production is so complicated that it seems the only way to achieve it would be to use the body as its own bioreactor. That is not to say that tissue engineering does not have potential. When large porous constructs are implanted, a limiting factor can be that blood vessels do not penetrate, even if the scaffold is bioactive and stimulating bone growth. Bone needs blood vessels if it is to survive, but also blood vessels will not grow into a porous construct if there is no living tissue inside it. This creates a bit of a 'chicken and egg' or 'Catch 22' situation, to which tissue engineering may be the answer. Stem cells could be grown in the pores of the scaffold in the laboratory prior to implantation so that, when the construct is implanted, there will already be viable cells present. The cell source has to be carefully considered. Perhaps an ideal cell source would be bone marrow stem cells from the patient. Bone marrow stem cells are thought to be those responsible for natural bone repair, where they differentiate into bone-producing cells (osteoblasts). Harvesting them from the patient is relatively straightforward compared to some other cell types, although the numbers that can be obtained are low. The aim would be to seed them within the scaffold and hope they attach and proliferate, and that the scaffold stimulates them to become bone cells rather than another cell type, such a cartilage cell or fibroblast.

Alternatively, blood vessels could be grown inside the scaffold in the laboratory, with the aim that they will connect up with existing vasculature. This strategy has been proven to work in mice.

Bone tissue engineering is not in widespread clinical use, unless you count mixing blood from the patient with a scaffold.

12.8 REGULATORY ISSUES

Once a promising new scaffold material has been developed and tested in the laboratory with cell culture tests and long-term mechanical tests as a function of degradation, it is time to translate the work from the bench to the bedside. This is not a trivial (or inexpensive) process. It takes considerable time and investment. Difficult decisions also have to be made. An important early decision is the class of material for which regulatory approval should be sought. At the time of writing, regulatory bodies, such as the FDA and the EU, are (quite rightly) making regulatory approval more difficult to achieve. Although it may disadvantage new companies with potentially important new products, safety must be paramount. The regulatory class depends on the claims made by a company. For example, in the EU, if a company claims that a scaffold will bond to bone, degrade over time and stimulate bone regeneration, it will be a Class 3 medical device. If the company claims a new material will simply do the same as other bone grafts and fill space, it will be a Class 2 device, but then the marketing or sales people in the company cannot claim performance superior to current products. Of course, to obtain regulatory approval of a Class 3 device takes more investment and may involve lengthy clinical trials. Claiming pharmaceutical properties of an implant may also mean that approval is needed from pharmaceutical (rather than device) regulatory bodies such as the UK Healthcare Products Regulatory Agency (MHRA).

12.9 SUMMARY

Bioactive glass foam scaffolds have the potential to improve performance of bioceramic bone grafts, but no bioceramics will ever fulfil all the criteria for an ideal scaffold. Inorganic–organic hybrids have the potential to be tough bioactive and biodegradable scaffolds. A massive challenge is expanding this technology to regenerate load-bearing skeletal components such as the hip. Replacing metals as load-bearing devices is the ultimate, but perhaps unachievable, goal. Tissue engineering approaches are not yet widely used in the clinic, except that surgeons often mix blood and bone marrow with implants, hoping to activate some stem cells already belonging to the patient. Incorporating osteoprogenitor cells within scaffolds may be the best solution if healthy bone is to be achieved in large defects.

FURTHER READING

Porous Bioceramics

Hing, K.A., Revell, P.A., Smith, N., and Buckland, T. (2006) Effect of silicon level on rate, quality and progression of bone healing within silicate-substituted porous hydroxyapatite scaffolds. *Biomaterials*, 27, 5014–5026.

Jones, J.R. and Hench, L.L. (2003) Regeneration of trabecular bone using porous ceramics. *Current Opinion in Solid State and Materials Science*, 7, 301–307.

Sepulveda, P., Binner, J.G.P., Rogero, S.O. *et al.* (2000) Production of porous hydroxyapatite by the gel-casting of foams and cytotoxic evaluation. *Journal of Biomedical Materials Research*, 50, 27–34.

Porous Melt-Derived Glasses

Elgayar, I., Aliev, A.E., Boccaccini, A.R., and Hill, R.G. (2005) Structural analysis of bioactive glasses. *Journal of Non-Crystalline Solids*, 351, 173–183.

Fu, Q., Rahaman, M.N., Bal, B.S. *et al.* (2008) Mechanical and *in vitro* performance of 13-93 bioactive glass scaffolds prepared by a polymer foam replication technique. *Acta Biomaterialia*, 4, 1854–1864.

Fu, Q., Saiz, E., and Tomsia, A.P. (2011) Bioinspired strong and highly porous glass scaffolds. *Advanced Functional Materials*, 21, 1058–1063.

Wu, Z.Y., Hill, R.G., Yue, S. *et al.* (2011) Melt-derived bioactive glass scaffolds by gel-cast foaming technique. *Acta Biomaterialia*, 7, 1807–1816.

Sol-Gel Foaming

Gough, J.E., Jones, J.R., and Hench, L.L. (2004) Nodule formation and mineralisation of human primary osteoblasts cultured on a porous bioactive glass scaffold. *Biomaterials*, 25, 2039–2046.

Jones, J.R., Ehrenfried, L.M., and Hench, L.L. (2006) Optimising bioactive glass scaffolds for bone tissue engineering. *Biomaterials*, **27**, 964–973.

Jones, J.R., Tsigkou, O., Coates, E.E. *et al.* (2007) Extracellular matrix formation and mineralization on a phosphate-free porous bioactive glass scaffold using primary human osteoblast (HOB) cells. *Biomaterials*, **28**, 1653–1663.

Sepulveda, P., Jones, J.R., and Hench, L.L. (2002) Bioactive sol-gel foams for tissue repair. *Journal of Biomedical Materials Research*, **59**, 340–348.

Porous Hybrids

Mahony, O., Tsigkou, O., Ionescu, C. *et al.* (2010) Silica–gelatin hybrids with tailorable degradation and mechanical properties for tissue regeneration. *Advanced Functional Materials*, **20**, 3835–3845.

Poologasundarampillai, G., Ionescu, C., Tsigkou, O. *et al.* (2010) Synthesis of bioactive class II poly(glutamic acid)/silica hybrids for bone tissue regeneration. *Journal of Materials Chemistry*, **40**, 8952–8961.

Valliant, E.M. and Jones, J.R. (2011) Towards softer materials for synthetic bone graft applications: hybrid materials. *Soft Matter*, **7**, 5083–5095.

Tissue Engineering of Blood Vessels Inside Scaffolds

Tsigkou, O., Pomerantseva, I., Spencer, J.A. *et al.* (2010) Engineered vascularized bone grafts. *Proceedings of the National Academy of Sciences*, **107**, 3311–3316.

13

Glasses for Radiotherapy

Delbert E. Day
Center for Bone and Tissue Repair, Graduate Center for Materials Research, Materials Science and Engineering Department, Missouri University of Science and Technology, Rolla, Missouri, USA

13.1 INTRODUCTION

This chapter is intended to give a general introduction, from the materials science perspective, of glasses that have been investigated, and in some cases are being used, for the *in situ* irradiation of diseased organs in the body or what is called brachytherapy. Radiation therapy of tumors works on the principle that radiation can damage the DNA in cells and, while turning cancerous, cells lose their ability to regenerate their DNA. Brachytherapy is not a new medical technique, but the use of glass as a radioactive delivery vehicle is relatively new, beginning about 25 years ago when radioactive glass microspheres were first used for the *in situ* irradiation of malignant tumors in the liver [1–3]. The types of glasses and their properties relevant to *in vivo* radiotherapy applications are described. Selected applications where glass microspheres have been used for *in situ* irradiation will also be described, particularly the treatment of patients with inoperable liver cancer.

Bio-Glasses: An Introduction, First Edition. Edited by Julian R. Jones and Alexis G. Clare.
© 2012 John Wiley & Sons, Ltd. Published 2012 by John Wiley & Sons, Ltd.

In general, either external beam or *in situ* radiation sources are used to irradiate diseased sites in the body. One of the most important advantages of *in situ* irradiation is that a weaker, shorter-range radiation can often be used and targeted to the specific tumor, and, consequently, damage to adjacent, healthy tissue is minimized. This means that larger doses of radiation can be safely used, which translates into a higher probability of destroying the malignant tumors.

From a theoretical standpoint, a wide range of radioisotopes emitting beta, gamma, or alpha radiation can be chemically incorporated into glass particles. Oxide glasses containing rare earth (RE) radioisotopes have been of great interest because the chemical and physical properties of these glasses, coupled with the properties of the RE radioisotopes, make them well suited for *in vivo* radiation therapy [4, 5].

Data for those RE glasses of general interest, and which are primarily beta emitters, are summarized in Table 13.1. Their high RE content, the half-life of the radioisotope, the range in soft tissue of the beta

Table 13.1 Selected properties[a] for rare earth (RE) aluminosilicate glasses and data for rare earth radioisotopes formed by neutron activation.

Property	Y_2O_3	Sm_2O_3	Ho_2O_3	Dy_2O_3
RE_2O_3 content (wt%)	35–55	32–64	36–69	47–65
Density (gm/cm^3)	2.80–3.89	3.31–4.65	3.39–5.55	4.07–4.99
Transformation temperature, T_g (°C)	885–895	780–820	860–878	856–874
Dilatometric softening point, T_d (°C)	935–945	815–875	898–912	882–910
Thermal expansion coefficient, α ($\times 10^{-7}$/°C)	31–70	51–75	41–60	52–66
Dissolution rate (gm/cm^2 min) in deionized water at 37 °C	1×10^{-9}	$1.6{-}30 \times 10^{-9}$	$0.5{-}3 \times 10^{-9}$	$<1 \times 10^{-9}$
Radioisotope	^{90}Y	^{153}Sm	^{166}Ho	^{165}Dy
Half-life (hours)	64	46.3	26.8	2.3
Range in tissue (mm)				
average	2.5	0.8	2	1.4
maximum	10.3	3.1	8.5	5.7
Maximum activity[b] (Ci/g)	2.4	61	63	230

[a]The lowest–highest values for a given property are shown.
[b]Per gram of radioactive nuclide in REAS glass, neutron flux of 10^{13} neutron/cm^{2}s.

radiation emitted by the RE radioisotope, and the ability to form these radioisotopes by neutron activation are all desirable features of these RE-containing glasses.

An oxide glass is a versatile material for the *in situ* delivery of therapeutic amounts of beta or gamma radiation. As we have seen in Chapter 2, by simple changes in the chemical composition, a glass can range from being bio-inert, to bioactive, to biodegradable in the body. During the melting process, most neutron activatable elements of interest, such as those in Table 13.1, easily dissolve in large amounts in the melt and become a strongly bonded, inherent part of the glass. As such, the radioisotope is confined to the glass, which minimizes, and in most cases prevents, the leakage of radiation from the target organ. In fact, borosilicate glasses are used to immobilize radioactive isotopes in nuclear waste. Glass can be made into several convenient shapes over a wide range in size, as shown by the microspheres and 1 mm diameter rods in Figure 13.1. In short, the versatility of glass is such that it is possible to design a biocompatible glass for a specific purpose and tailored to a particular target organ in terms of the radioisotope that it contains (type of radiation and half-life), its biodegradability (or lack thereof), and its shape and size. Glass compositions can be tailored as to whether they are degradable or not, depending on the need and application.

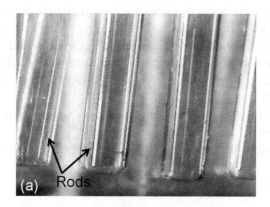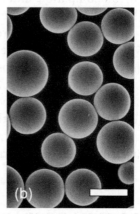

Figure 13.1 Examples of (a) rods and (b) microspheres made from rare earth aluminosilicate glasses for use as *in situ* radiation delivery systems (brachytherapy). The 1 mm diameter rods in (a) and the glass microspheres in (b) have a nominal composition of 46.8 wt% Sm_2O_3, 18.2 wt% Al_2O_3, and 35.0 wt% SiO_2, and 55 wt% Y_2O_3, 20 wt% Al_2O_3, and 25 wt% SiO_2, respectively. The white bar in (b) is 10 μm.

13.2 GLASS DESIGN AND SYNTHESIS

There are two general ways to make a radioactive glass for *in vivo* use. The first is to add a radioactive substance to the batch materials and to melt these materials in the normal way so that the radioisotope dissolves in the melt and becomes an integral part of the glass. This method has been used successfully, but the special precautions that must be taken to handle radioactive materials during the melting and subsequent processing is a serious disadvantage of this method.

The second method is to melt a glass in the normal fashion using non-radioactive materials and to then make the glass radioactive by neutron activation as the last step in the fabrication process. This greatly simplifies the melting and subsequent handling of the glass, because it is non-radioactive and no special precautions or procedures are required before the neutron activation step. However, the chemical composition of the glass is limited in this instance, because many of the elements commonly present in oxide glass, such as Na, K, and Ca, cannot be present as they would also become radioactive by neutron activation.

A glass used for the *in situ* irradiation of a target organ and which will undergo neutron activation should possess the following five general characteristics [5]. It should be:

(a) biocompatible and non-toxic in the body;
(b) chemically resistant to the body fluids to the extent that none of the radioisotope is released in the body, and after the radioisotope has decayed, then release can occur if biodegradability is desired;
(c) made with a composition sufficiently high in a neutron activatable element so that the level of specific activity needed for the desired treatment is achieved;
(d) free of any element that will form an unwanted radioisotope during neutron activation; and
(e) capable of being formed into particles or spheres of the desired size and shape.

13.3 NON-DEGRADABLE OR BIO-INERT GLASSES: RARE EARTH ALUMINOSILICATE GLASSES

Many applications require that the glass remains stable in the body and does not degrade. The composition therefore has to be designed to be resistant to corrosion in body fluid.

13.3.1 Preparation

Several families of rare earth aluminosilicate (REAS) glasses possess the properties listed in Section 13.2 and have been investigated [4–7] for the *in situ* irradiation of diseased organs. These glasses have a simple chemical composition, being composed of just three oxides, namely, alumina (Al_2O_3), silica (SiO_2), and the desired, neutron activatable rare earth oxide (RE_2O_3). Unlike most common glasses, these glasses contain only four elements, Al, Si, O, and the RE. Fortunately, the radioisotopes formed from Al, Si, and O during neutron activation decay rapidly and are of no consequence. These glasses will hereafter be designated as REAS, where RE designates the rare earth, A denotes alumina, and S denotes silica. Thus, a REAS glass containing samarium (a RE element) would be designated as SmAS.

REAS glasses are prepared by mixing high-purity powders of the three oxides and melting the homogeneous mixture in a platinum crucible for 4–8 hours at 1500–1600 °C. Care must be taken in choosing the raw materials to ensure that they are of high chemical purity and do not contain traces of unwanted neutron activatable elements.

The glass-forming region for several families of REAS glasses, which can be melted below 1600 °C, is shown in Figure 13.2, which is a similar style of diagram to Figure 2.1, the composition map of bioactive glasses.

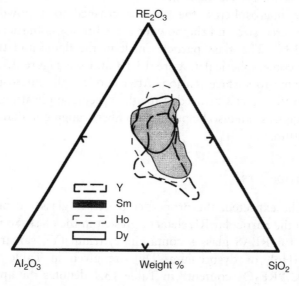

Figure 13.2 Compositional diagram showing the glass formation range for rare earth aluminosilicate (REAS) compositions that melt below 1600 °C [4, 6, 7].

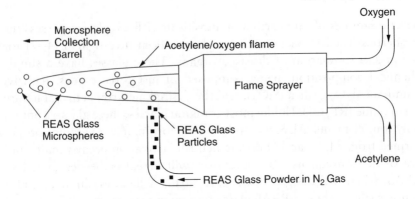

Figure 13.3 Schematic of the system used to produce glass microspheres by spheroidizing a glass powder in a gas burner (flame spheroidization). (Reprinted with permission from [4]. Copyright (1994) Trans Tech Publications.)

REAS glasses form over a fairly broad compositional range in each system, which is a desirable feature. Note also that these glasses can contain large amounts of the various RE oxides, from a low of 32 wt% to a high of 69 wt%, depending upon the specific RE oxide. Yttrium, while not strictly a RE, is included in Figure 13.2 since its properties are close to those for the RE elements.

To make microspheres, the glass is crushed to a powder of the generally desired size and the powder is feed to a gas burner as shown in Figure 13.3. The glass particles melt in the flame and the molten droplets become spherical (Figure 13.1b and see Figure 15 in colour section) owing to surface tension. After cooling, the microspheres are screened to the exact size needed for the specific application. Many of the REAS glasses can also be pulled into fibers ranging in diameter from 10 to 5000 μm.

13.3.2 Properties

As would be expected, the properties of REAS glasses depend somewhat upon the particular RE element present in the glass. Some selected properties for REAS glasses containing yttrium (Y), samarium (Sm), holmium (Ho), or dysprosium (Dy) are given in Table 13.1. The line labeled "RE_2O_3 content" in Table 13.1 denotes the approximate minimum–maximum amount of RE oxide for each system. The density of the REAS glasses in Table 13.1 increases with increasing RE_2O_3

concentration, and is typically two to four times higher than that of blood. Additional property data for these and other REAS glasses can be found elsewhere [4, 6, 8].

An important property of REAS glass for *in vivo* use is their outstanding chemical durability. Since the radioisotope is a chemical and physical part of the glass structure, as opposed to a surface coating, the radioisotope can only escape from a target organ if the glass dissolves in the body fluids during the time it is radioactive.

Because of their excellent chemical durability in the biological environment, the release or leakage of the radioisotope from a REAS glass is so small as to be physiologically unimportant, in most applications. Several thousand patients with primary liver cancer have been treated with radioactive yttrium aluminosilicate (YAS) glass microspheres (40 wt% yttria) for over 20 years with no reported incident of excessive leakage of the ^{90}Y from the liver due to the dissolution of the YAS glass.

13.4 BIODEGRADABLE GLASSES: RARE EARTH BORATE/BOROSILICATE GLASSES

There are certain biological applications, such as radiation synovectomy of arthritic joints, where it is desired for the glass delivery vehicle to degrade gradually in the body once it is no longer radioactive. While not in commercial use at this time, alkali borate (B_2O_3) and borosilicate glasses containing RE oxides, along with other cations, are known to be biodegradable [9, 10]. Some typical compositions of such glasses are given in Table 13.2 and more compositions can be found in Refs [11, 12].

Table 13.2 Composition (wt%) of selected biodegradable glasses and their reaction rate in phosphate-buffered saline (PBS) solution.

Oxide	DyLB3-10	HoLB3-10	DyLB3-30	DyLAB5	DyLAB10	LS-Y
Li_2O	11.3	11.3	8.8	6.1	5.6	13.7
B_2O_3	78.7	78.7	61.2	58.4	53.4	–
Dy_2O_3	10.0	–	30.0	23.7	23.1	–
Al_2O_3	–	–	–	6.5	12.6	–
MgO	–	–	–	1.5	1.5	–
SiO_2	–	–	–	3.8	3.7	59.2
Y_2O_3	–	–	–	–	–	11.4
Ho_2O_3	–	10.0	–	–	–	–
Reaction rate[a] (μm/h)	6 (37 °C)	–	0.3 (21 °C)	0.07 (22 °C) 0.4 (42 °C)	–	11 (37 °C)

[a]The rate at which a surface layer develops on the surface of a microsphere when it is immersed in PBS solution, pH = 7.4, at the temperature indicated [9, 10].

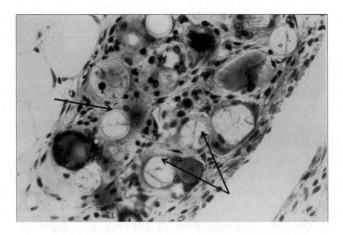

Figure 13.4 DyLB3-10 glass microspheres embedded in the synovial tissue of the stifle joint of a rat. Microspheres are shown after two weeks in the stifle joint and degradation of the microspheres has started as indicated by the cracks visible in the microspheres marked with the arrows. (Adapted with permission from [9]. Copyright (2002) John Wiley and Sons, Inc.)

Unlike the REAS glasses, which have such a high chemical durability that they are expected to remain in the body for long periods of time (years), the biodegradable glasses have a much lower chemical durability with respect to the body fluids and begin to biodegrade in a matter of hours to weeks (see Figure 13.4). The biodegradation process may start even when the glass is radioactive as long as the radioisotope does not escape from the target site. A somewhat surprising feature of these glasses is that *in vitro* and *in vivo* studies [9, 12] show that, even though the glass is dissolving or degrading in the body, the radioisotope of RE elements such as Dy reacts with the phosphate and other anions in the body fluids to form an insoluble phosphate material, which confines the radioisotope to the target site.

The rate at which these borate glasses react with the body fluids and undergo biodegradation is known to depend upon the overall glass composition (Table 13.2) and is likely to depend on the specific biological environment at the target site, although there is no evidence of the latter at this time. Chapter 6 discusses borate glasses of even higher solubility for tissue regeneration applications. As is evident from Table 13.2, a higher borate (B_2O_3) content increases the biodegradation rate, while a higher RE_2O_3 content and the addition of oxides such as Al_2O_3 and SiO_2 to the glass reduces the biodegradation rate.

Biodegradable alkali borate glasses should possess the same five general properties as mentioned above for REAS glasses. While borate glasses are designed to be less chemically durable than the REAS glasses, it is still important that the radioisotope not be released and escape from the target site during the degradation process. Since it is expected that these glasses will also undergo neutron activation to form the radioisotope of choice, it is important that the glass be free of any trace impurities that would form unwanted radioisotopes.

There may be concern about irradiating a glass in a nuclear reactor that contains boron, because ^{10}B has a high cross-section for neutrons. In most instances, the amount of borate glass being irradiated is usually small (<100 g), so the total amount of ^{10}B is normally not a problem. However, if needed, the glass can be made using B_2O_3 enriched with ^{11}B, which eliminates the high neutron absorber ^{10}B.

Similarly, there are instances where it is desired for the borate glass to contain Li_2O, but during neutron activation the naturally occurring ^{6}Li forms radioactive tritium, ^{3}H. This situation can be avoided by using a raw material enriched in ^{7}Li.

13.5 DESIGN OF RADIOACTIVE GLASS MICROSPHERES FOR *IN VIVO* APPLICATIONS

Before discussing specific applications where glass particles are used for *in situ* irradiation purposes, some general aspects of this type of treatment that are common to different *in vivo* applications will be described.

13.5.1 Glass Particle Shape

Commonly, a glass radiation delivery vehicle is in the shape of a microsphere ranging from 10 to 30 μm in diameter (Figure 13.1b), although glass fibers or seeds and even irregular shapes could also be used in some applications. As shown in Figure 13.1, glass microspheres have a smooth surface and their size and uniformity can be carefully controlled and even tailored (sized) for a particular organ. A spherical shape is also desirable when the procedure is either to inject the microspheres into the blood stream and let the blood deliver the microspheres to the target site (liver) or to inject the microspheres directly into the target site (radiation synovectomy of a joint). The microspheres are sized so that they will enter the capillary bed of a target organ, but are too large to pass completely through the organ. Thus, the target organ acts like a filter in which the radioactive glass particle becomes trapped.

13.5.2 Useful Radioisotopes

An important advantage of *in situ* irradiation is that much larger, more than 10 times, therapeutic doses of radiation can be delivered using radioisotopes that emit the more localized (shorter-range) beta radiation. In comparison, radiation administered externally has to pass through other tissue, and shaped radiation beams have to be aimed from several angles, meeting at the tumor in order to reduce the dose in the surrounding, healthy tissue. To date, REAS and borate glass microspheres containing beta-emitting ^{90}Y, ^{153}Sm, ^{165}Dy, ^{166}Ho, and $^{186}Re/^{188}Re$ have been tested in animals [5–7, 9, 10, 13, 14]. Only REAS microspheres containing ^{90}Y are in commercial use (TheraSphere™) to treat patients with primary liver cancer, hepatocellular carcinoma (HCC). In addition to the radioisotopes mentioned above, other radioisotopes potentially useful for *in situ* irradiation are ^{32}P, ^{108}Pd, ^{113}In, ^{124}Te, ^{146}Nd, ^{168}Yb, and ^{177}Lu.

13.5.3 Radiation Dose

The maximum amount of radiation that glass microspheres can safely deliver to target organs is still undetermined, but should be expected to depend upon such factors as the specific organ being irradiated, the specific radioisotope and type of radiation emitted, and the specific activity of the microsphere at the time of injection. In general, the radiation doses are much larger than those administered by an external beam. For example, prior to use in humans, dogs were injected with ^{90}YAS glass microspheres that delivered up to 350 Gy (35 000 rad) to their liver with no ill effects [15]. Patients with inoperable liver cancer (HCC) are now being injected with YAS microspheres, <100 mg, that deliver a dose calculated to be up to 150 Gy (15 000 rad) to the entire liver [16, 17]. Since the glass microspheres tend to locate preferentially in the malignant tumor(s), as opposed to the healthy tissue, localized regions in the tumor have received doses up to 3000 Gy (300 000 rad), as calculated from the number and distribution of YAS glass microspheres found in a treated liver [18].

In the early human trials, it was anticipated that a patient would receive a single injection of beta-emitting radioactive microspheres, because there was little information available for judging the effect of such large doses on the target organ. However, the early patients with liver cancer tolerated doses up to 150 Gy quite well and, as confidence in the safety of these large doses has increased, an increasing number of patients have received repeated injections (two to four) of the ^{90}Y microspheres, with no reported ill effects. In one case, a male with inoperable liver cancer received eight injections of ^{90}Y microspheres (at 150 Gy per injection) over a four-year period with no ill effects.

13.5.4 Tumor Response and Tailoring of Glass Composition

The response of tumors receiving *in situ* beta irradiation from REAS glass has been demonstrated in animal experiments [19, 20] as well as in humans [2, 3, 16, 21]. In an animal experiment [19], the growth (or lack thereof) was determined for tumors in six nude mice that had been injected with approximately 2×10^7 BT-20 cells (human mammary carcinoma). The tumor in each mouse was allowed to grow for 14 days, at which time the tumor was injected directly with either 5 mg of radioactive $^{166}HoAS$ glass particles ($\sim 200\,\mu Ci$/tumor) or 5 mg of non-radioactive HoAS glass particles. The volume of the tumor was measured at the time it was injected with the HoAS glass particles and 12 days later.

As shown in Figure 13.5, the volume of the tumor for each animal (numbers 1, 2, and 3) in the control group, injected with non-radioactive HoAS glass, continued to grow significantly, as expected. However, the volume of the tumors injected with the radioactive $^{166}HoAS$ glass particles, either decreased, animals 4 and 5, or remained the same, animal 6.

In the second experiment [20], varying amounts of radioactive ^{90}YAS glass microspheres (20–25 μm in diameter) were injected directly into a tumor (produced from SMM C-7721 cells) that had been induced in nude mice. As is evident from Figure 13.6, the tumor growth that occurred in each animal during the 14 days after injection was significantly less with increasing quantity of radioactive ^{90}YAS glass microspheres.

These two examples show that the *in situ* beta radiation from REAS glass containing ^{166}Ho or ^{90}Y is effective in not only preventing the

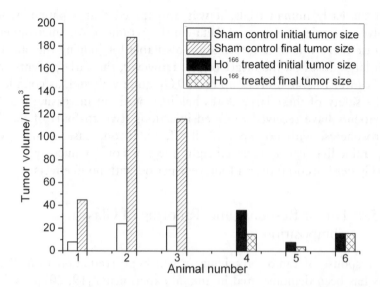

Figure 13.5 Behavior of BT-20 tumor xenografts in six nude mice. Initial tumor volume (solid bars) was measured 14 days after animals were injected with BT-20 cells. The tumor in each animal was then injected with either 5 mg of 2–5 μm glass particles (animals 1–3) or radioactive (^{166}Ho) particles (animals 4–6). The glass composition was 7 wt% Ho_2O_3, 20 wt% MgO, 21 wt% Al_2O_3, and 52 wt% SiO_2. The final tumor volume (hatched bars) was measured 12 days after injecting the glass particles. (Drawn from data presented in Ref. [19].)

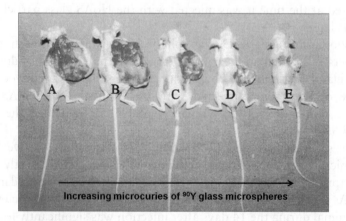

Figure 13.6 Example of how the growth of a malignant tumor is reduced when injected with ^{90}YAS glass microspheres. Tumor shown for each mouse is 14 days after each mouse was injected with zero (A), 10 (B), 50 (C), 100 (D), and 150 (E) microcuries of ^{90}Y glass microspheres. (Adapted from [20]. Copyright (1998) Digest of Chinese Academic Journals.)

growth of a tumor, but also reducing the size of an existing tumor. The overall effectiveness of *in situ* irradiation should be expected to depend upon many factors, such as the characteristics (vascularity) of a specific tumor, the energy and range of the radiation, the half-life of the beta-emitting radioisotope, and the amount of radioactive isotope inside the tumor(s). Since REAS glasses containing a preferred beta-emitting radioisotope can be easily prepared, it is possible to select the most effective beta-emitting radioisotope for treating a particular type of tumor in a specific organ. In other words, it is possible to tailor-make an *in situ* radiation delivery vehicle that is optimized for a specific organ or type of tumor.

In the work undertaken to date, only REAS glasses containing a single radioactive element have been used for the *in situ* irradiation of diseased organs. However, REAS glasses containing two or more neutron activatable RE elements are easily prepared, so a target organ could be irradiated simultaneously with radiation emitted from several radioisotopes for enhanced effectiveness. Combining several neutron activatable radioisotopes of different half-lives and emission energies into a single glass could be advantageous in optimizing the irradiation of tumors of different size and in delivering the radiation dose over a time period not attainable with a single radioisotope. Another alternative would be to blend two or more REAS glasses, each of which contains a different neutron activatable element, so that the delivered dose consists of a combination of radiation type, energy, and half-life.

13.6 TREATMENT OF LIVER CANCER: HEPATOCELLULAR CARCINOMA

Liver cancer (hepatocellular carcinoma) is difficult to treat since there are commonly few symptoms until the life of the patient is in peril. Approximately 18 500 primary tumors of the hepatic and biliary tree are diagnosed each year, which result in an estimated 12 300 deaths [13]. The number of metastatic liver tumors diagnosed each year is estimated to be roughly 10 times the number of primary tumors.

At the time of diagnosis, surgery is often not an option because of the presence of multiple tumors and the difficulty in removing them completely. External beam radiation is generally ineffective since the maximum dose that can be delivered is limited by the damage to nearby healthy tissue and is typically too small to destroy the tumor(s). Chemotherapy by itself usually provides only temporary improvement.

The advantages of irradiating hepatic tumors with an *in situ* radiation source so as to avoid damaging nearby healthy tissue and thereby deliver larger doses (>100 Gy) was recognized in the early 1960s when patients were treated with ceramic microspheres, 40–60 μm in diameter, which contained the beta emitter ^{90}Y [22–24]. These early studies were not without problems, such as leakage of the radioisotope from the tumor site and the migration of radioactive particles to other organs. Nevertheless, the procedure of infusing radioactive particles into the liver via the hepatic artery was deemed to be simple and reasonably safe, reductions in tumor size were noted, complications were minimal, and clinical and subjective improvements were noted in some patients. The remainder of this section will focus on the use of radioactive ^{90}YAS glass microspheres to treat patients with primary liver tumors, HCC, that started in the mid-1980s. In 2000, YAS glass microspheres were approved by the US Food and Drug Administration (FDA) for treating patients with inoperable primary liver cancer (HCC).

Patients were first treated with ^{90}YAS glass microspheres at the University of Michigan [15, 21] and later at Toronto General Hospital in Canada [3] in dose escalation studies. The microspheres were made from YAS glass, which had been selected for its outstanding chemical durability and high yttria content. Yttrium was the element of choice, since ^{90}Y emits beta radiation, which has an average range of 2.5 mm in soft tissue, has an acceptable half-life, and can be formed by neutron activation of the naturally occurring ^{89}Y, which is 100% abundant.

The technique employed to deliver the radioactive microspheres to the liver via the hepatic artery is depicted schematically in Figure 13.7. A catheter is inserted into the patient and manipulated to the desired location. The desired amount (typically <100 mg) of 20–30 μm radioactive ^{90}YAS microspheres (Figure 13.1) is flushed from the dose vial with a saline solution into the catheter, where the microspheres enter the blood stream and are eventually deposited in the capillary bed of the targeted tumor(s). The infusion of radioactive microspheres to either a single or multiple tumors is complete in a few minutes.

The results of the early studies were promising, and doses up to 150 Gy, calculated as whole liver dose, were deemed safe. Since the ^{90}YAS glass microspheres tend to follow the blood flow, the microspheres preferentially concentrate in the tumor mass at ratios from 10 to 40 times higher than in the healthy tissue. Because of the outstanding chemical durability of the YAS glass, no leakage of the ^{90}Y from the target site was detected, and no serious side effects were observed. Over time, the infusion techniques have improved, especially the positioning of

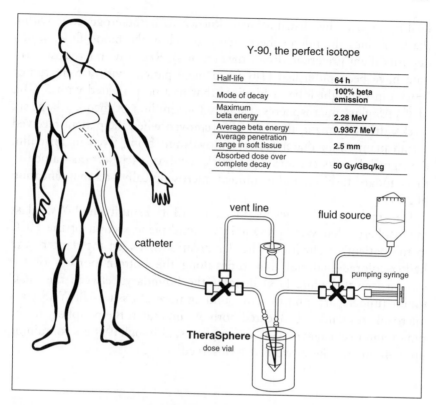

Figure 13.7 Schematic of the delivery of radioactive yttrium aluminosilicate (YAS) glass microspheres to malignant tumors in the liver and data for the Y-90 radioisotope. (Reprinted with permission from Nordion, Ottawa, Canada. Copyright (2012) Nordion.)

the catheter, so that microspheres are now infused selectively into either lobe of the liver and a larger fraction of the radioactive microspheres are delivered to the specific target tumor(s) as opposed to the healthy liver tissue [16]. A typical dose for delivering 150 Gy to the whole liver contains from four to six million glass microspheres with an initial activity of ~2500 Bq/sphere.

As of 2011, approximately 6000 patients with primary liver cancer have been treated with ^{90}YAS glass microspheres, some receiving multiple injections as mentioned previously. Patients are treated on an outpatient basis and side effects are minimal in most cases, so the patients' quality of life remains high.

There has been much interest in gaining a better picture of the distribution of the YAS microspheres within the tumor and surrounding healthy tissue. This distribution is known to be heterogeneous within the liver since the microspheres tend to follow the blood flow, which deposits them preferentially in the tumor(s). Recently, microdosimetry data have been published [18] for a male patient with a 5 cm tumor (HCC) in the right lobe of his liver that had been treated with 5 GBq of ^{90}YAS glass microspheres (roughly four million of them). About six weeks after treatment, the liver was removed when the patient received a transplant. Immediately after removal and being fixed in formalin, the right lobe was cut into 8 μm thick sections, each section stained in the standard fashion, and examined microscopically at a magnification of 200×.

The small dark dots visible in Figure 13.8 are individual YAS glass microspheres that were present in the mid-plane section of the HCC in the patient's right lobe. The demarcation between the darker- and lighter-colored material that runs along the bottom portion of the section (see the white box) denotes the boundary between the HCC tumor (upper part) and the surrounding normal tissue. The YAS glass microspheres tend to be located both within and at the periphery of the tumor and in clusters of 1–4 microspheres, although clusters containing up to 40 microspheres were also observed.

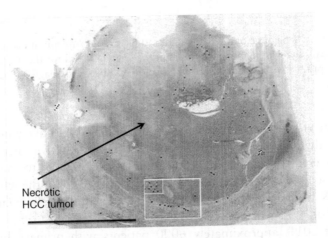

Figure 13.8 Tissue sample from the liver of a patient with a unresectable hepato-cellular carcinoma (HCC) 6 weeks after the patient had been treated with 5 GBq of ^{90}Y aluminosilicate glass microspheres (typically four million of them). The small dark dots, which are confined primarily within the tumor (darker portion of the section) and at the tumor–healthy tissue interface, are individual glass micro-spheres, 25 ± 10 μm. The bar in the lower left corner represents 1 cm. (Adapted with permission from [18]. Copyright (2004) Elsevier Ltd.)

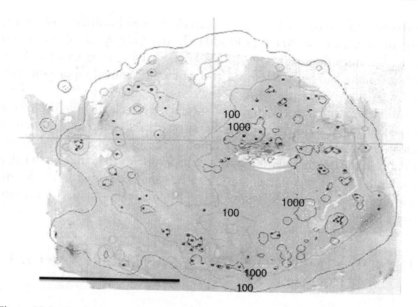

Figure 13.9 Isodose curves overlaid on the histology section shown in Figure 13.8. The pale gray and darker gray curves denote a dose of 100 and 1000 Gy, respectively. The dose curves were calculated from the entire tissue section, which contained 852 of the YAS glass microspheres with a nominal activity of 2500 Bq/sphere at the time of injection. Calculations indicated that localized regions received doses up to 8000 Gy. The dose rapidly decreased at the normal tissue interface, outer curve. The bar in the lower left corner represents 1 cm. (Adapted with permission from [18]. Copyright (2004) Elsevier Ltd.)

The isodose curves (showing lines of same dose) shown in Figure 13.9 were calculated from a 3D dose grid of the entire tissue section, which contained 852 glass microspheres whose computed nominal activity was 2500 Bq/microsphere at the time of infusion. The pale gray and darker gray curves denote a dose of 100 Gy (10 000 rad) and 1000 Gy (100 000 rad), respectively. The isodose curves in Figure 13.9 were calculated using the activity of each microsphere, the decay constant for ^{90}Y, and the position of each microsphere within the tissue section (and the two adjacent sections).

By counting the number of microspheres in the tumor and adjacent normal tissue in the section in Figure 13.8, the ratio of the number of microspheres in the tumor to normal tissue was found to be 16 : 1. In addition, the isodose curves demonstrate that a major fraction of the total dose was confined within the tumor, and the dose rapidly decreased within about 4 mm of reaching the tumor–normal tissue interface. Even within the tumor itself, the dose is localized to small regions and can be massive. Dose–volume histogram analysis showed localized doses in the tumor as high as 8000 Gy (800 000 rad), but only 2.5% of the tumor received more that 1000 Gy (100 000 rad). It was estimated that 85% of the tumor had been destroyed at the time of removal (approximately six weeks).

Additional microdosimetry data should give a clearer picture of the expected dose delivered to specific sites, thereby making possible a more accurate assessment of the amount of radiation delivered to the tumor(s) and to the healthy part of the liver. Ideally, this knowledge could mean that even larger doses can be safely delivered in the future without exceeding the radiation limit of 20–35 Gy (3500 rad) for healthy liver tissue.

13.7 TREATMENT OF KIDNEY CANCER: RENAL CELL CARCINOMA

When discovered at an early stage, surgical removal of a diseased kidney is the usual treatment of choice for patients with kidney cancer (renal cell carcinoma). However, in cases where a malignant tumor is at an advanced stage or has spread to other nearby organs, surgery may not be possible owing to the risk that renal cancer cells will remain in the patient and malignant tumors will reappear elsewhere, metastatic tumors. In this case, treatment options are limited and generally unsatisfactory.

An alternative treatment for patients with inoperable kidney cancer is to infuse the cancerous kidney/tumor with radioactive glass microspheres for the purpose of quickly destroying the cancer cells in the kidney and malignant tumor. Once the tumor and cancer cells are destroyed, and the risk of viable cancer cells remaining in the body is greatly diminished, the diseased kidney can be surgically removed.

The idea of destroying a diseased kidney prior to surgical removal has been investigated in rabbits [25] using glass microspheres (20–40 μm in diameter) made from a chemically durable glass that contained 7 wt% Sm_2O_3, 20 wt% MgO, 21 wt% Al_2O_3, and 52 wt% SiO_2. This glass was made radioactive by neutron activation to form [153]Sm, which is primarily a beta emitter with an imageable gamma ray (Table 13.1). In *in vivo* trials, using the same general technique as shown in Figure 13.7, varying amounts of radioactive [153]Sm microspheres were injected into the renal artery and deposited into one kidney of 18 healthy New Zealand White rabbits. In this dose escalation experiment, the target kidney received from 16 to 266 MBq of [153]Sm, corresponding to doses ranging from 5 to 400 Gy (500–40 000 rad). The rabbits were sacrificed over a period of 14 months, during which time there were no deaths due to radiation-related complications.

Histological examination of both kidneys, as well as other organs, revealed no damage of significance except to the kidney in which the ^{153}Sm microspheres had been injected. The only change in nearby organs was a slight scarring of the fat tissue adjacent to the target kidney. On the other hand, the degree of histological damage to the target kidney was significant and increased with increasing dose. At the time it was surgically removed, the size of the irradiated (150 Gy) kidney was only 25–35% of that of the control (non-injected) kidney.

It is encouraging that none of the rabbits showed any outward sign of radiation effects, even at the highest dose, and the only organ showing any detectable histological change was the kidney injected with the radioactive ^{153}SmAS microspheres. The favorable histological evidence from this study indicates that this *in situ*, highly localized radiation treatment of inoperable tumors in the kidney offers the possibility for sterilizing such tumors prior to their surgical removal.

13.8 TREATMENT OF RHEUMATOID ARTHRITIS: RADIATION SYNOVECTOMY

Rheumatoid arthritis is a chronic inflammatory disease that can cause destruction of articular cartilage in joints and can also affect the lungs and other tissues. Primarily an inflammatory response occurs in the synovial membrane (synovial tissue) of joints, which is the soft tissue encapsulating the synovial fluid and the joint.

The injection of radioactive substances into an arthritic joint for the purpose of reducing the swelling and inflammation of the synovial tissue, a procedure commonly referred to as radiation synovectomy, is an established procedure in Europe. A schematic where biodegradable radioactive glass microspheres or particles are used is shown in Figure 13.10. In this treatment, particles containing radioisotopes such as ^{90}Y, ^{198}Au, ^{186}Re, and ^{169}Er and ranging in size from a few nanometers (colloids) to 5–15 μm (one to three times the diameter of a red blood cell) are injected into the affected joint [26].

In this application, it is considered desirable for the particles to eventually degrade *in situ* or otherwise be eliminated from the joint being treated. Encouraging results have been obtained from animal experiments where the stifle (knee) joint of rabbits and rats have been

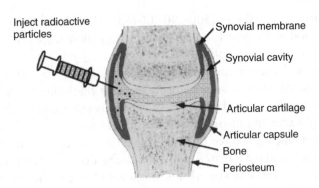

Figure 13.10 Concept of using biodegradable radioactive glass microspheres for radiation synovectomy of rheumatoid arthritic joints. It is important that no radioactivity "leaks" from the treated joint. (Modified from Ref. [27].)

injected with radioactive and non-radioactive microspheres, 5–15 μm in diameter, made from lithium borate [9, 28] or lithium silicate glass [29]. In addition to degrading in the animal, the leakage of the radioisotope, either ^{165}Dy or ^{90}Y, from the joint was much less than the 10–25% observed when radiocolloids are used. Furthermore, no evidence has been found that the glass microspheres cause any mechanical wear or damage in the joint for periods up to six months.

Microspheres made from a dysprosium–lithium borate glass (DyLB3-10 in Table 13.2) have been tested [9] for up to 64 days in a simulated synovial fluid (SSF) at 37 °C and in the stifle joint of Sprague Dawley rats [9] for 112 days. The SSF was a commercial phosphate-buffered saline solution containing 0.3 wt% hyaluronic acid at pH = 7.4. Animals were sacrificed after 28 days for histological examination of the stifle joint and at 112 days to determine the condition of the microspheres, by scanning electron microscopy (SEM), and the general condition of the joint.

Chemical analysis of the SSF solutions showed that the *in vitro* degradation of non-radioactive DyLB3-10 microspheres started within 6 hours, and within 12 hours the concentration of Li, B, and Dy in solution had become nearly constant. After 64 days, the SSF contained approximately 97% of the Li and 80% of the B in the initial glass, but only 0.3% of the Dy in the microspheres had dissolved. Even though the starting glass had nearly dissolved, 70–75% weight loss, only 0.3% the Dy in the microsphere had been released to the solution.

Additional degradation experiments with radioactive DyLB3-10 microspheres, neutron activated to form the beta-emitting ^{165}Dy (half-life 2.3 hours), verified that very little of the radioactive ^{165}Dy was released during the degradation of this borate glass in SSF. After being immersed in the SSF for 7.2 hours (or three half-lives for ^{165}Dy), which corresponds to the decay of 87.5% of the initial activity, less than 0.1% of the ^{165}Dy initially present in the microspheres was detected in the SSF. These results for the DyLB3-10 glass indicate that leakage of the radioactivity during the degradation of the microspheres should be of minimal concern.

Similarly encouraging results were obtained from *in vivo* experiments [9] in which 1 mg of non-radioactive DyLB3-10 microspheres, from 5 to 15 μm in diameter, were injected in the stifle joint of a rat. The typical appearance of the remnants of DyLB3-10 microspheres after being in the rat stifle joint for 14 days is shown in Figure 13.4. While degradation of the microspheres clearly occurred, the diameter of the remnants is close to that of the original microspheres. Energy-dispersive X-ray spectroscopy (EDS) analysis of several remnants showed that they consisted primarily of dysprosium along with phosphorus and calcium from reaction with the body or synovial fluids. Most likely, this dysprosium phosphate-rich material was formed as the glass microsphere degraded such that the dysprosium reacted with the body fluids, which is the source of Ca and P.

Histological examination of the stifle joints revealed that the glass microspheres were embedded within the synovial membrane [9, 10]. No microspheres were found in the articular cartilage and there was no evidence of mechanical damage to joint tissue. Typically, clusters of microspheres were observed dispersed throughout the synovial membrane. Tissue response was mild and no serious inflammation was observed. The only foreign-body response observed consisted of macrophage/giant cell formation and the proliferation of fibrous tissue around the agglomerated microspheres. This is normal for anything implanted that is biocompatible but not bioactive.

Based on the *in vitro* and *in vivo* results, the DyLB3-10 borate glass microspheres were considered good candidates for radiation synovectomy [9]. The microspheres degraded *in vivo* at an acceptable rate and the leakage of radioactivity from the joint was quite small, 0.1–0.2%, which is much less than the materials now being used for radiation synovectomy in Europe. Other than mild inflammation, there was no physical or histological damage to the tissue within a joint. The DyLB3-10 glass microspheres were easily neutron activated: 1 hour at a neutron flux of 8×10^{13} cm^{-2} s^{-1} yielded a specific activity of ^{165}Dy of 220 Ci/g, which should be adequate for most synovectomy treatments.

In another study [29], microspheres, 5–15 μm in diameter, of the lithium silicate glass (LS-Y) listed in Table 13.2 and containing radioactive ^{90}Y (half-life of 64.2 hours) were injected into the stifle joint of 16 New Zealand White rabbits that had been treated with antigens to induce arthritis in the stifle joint being treated. After neutron activation of the glass microspheres to produce ^{90}Y, the microspheres were suspended in a sterile saline solution containing 1% hyaluronate and injected into the desired stifle joint. The mean dose injected for low-activity microspheres, 50 μCi/mg, and high-activity microspheres, 500 μCi/mg, was 158 and 351 μCi, respectively.

Leakage of the radioactivity from the joint one week after injection, as determined from the activity of nine organs and excreta, varied from a low of 0.28% to a high of 0.82% of the injected dose over seven animals. These values are comparable to the 0.1% Y that was leached from this glass after being in saline solution, 37 °C, for six weeks. This amount of leakage is much less than that reported for other materials now in use. It should be noted that the small amounts of tritium formed from the ^{6}Li in these glasses during neutron activation could be avoided by using a glass that contains only non-activatable ^{7}Li.

Histological examination of the whole joint collected at one week after injection showed that clusters of 5–10 microspheres were embedded 50–100 μm deep in the synovial lining, similar to what was observed when DyLB3-10 glass microspheres were injected into the stifle joint of rats [9]. These results indicate that the glass microspheres, which number from 400 000 to 3–5 million depending upon their diameter and the dose being administered, do not remain suspended in the synovial fluid, but instead become embedded in the synovial tissue after some unknown period of time, less than one week, where they remain. After three weeks, the microspheres were enveloped by cells and the average diameter of the microspheres had not changed.

After one week *in vivo*, some of the microspheres were surrounded by a halo suggestive of slow degradation. The diameter of the treated joints, which was used as a measure of joint inflammation, remained constant within experimental error. A partial explanation for the lack of improvement may be the non-uniform distribution of the microspheres in the synovial lining, which could permit the disease to progress in those parts of the joint where microspheres were absent.

The overall results of these two studies indicate that biodegradable glass microspheres are candidates for radiation synovectomy purposes. In both studies, the release or leakage of radiation from the joint was less than 1% and within tolerable limits even for the rapidly degrading

DyLB3-10 glass. In this instance, the larger size of the glass microspheres compared to colloidal particles is believed to be an advantage in reducing the leakage of radioactivity from the joint. Another promising feature is that the glass microspheres become embedded and immobilized in the synovial lining fairly rapidly such that they do not cause any detectable amount of physical or mechanical damage to the joint tissue. It is also clear that therapeutic doses of radiation can be delivered by a few milligrams of neutron activated, radioactive glass microspheres. On the other hand, a more uniform dispersion and distribution of the microspheres throughout the synovial lining is desirable to achieve improved therapeutic results.

13.9 SUMMARY

An expanding number of investigations have shown that glass microspheres are effective as *in situ* radiation delivery vehicles for treating diseases such as cancer and rheumatoid arthritis. The versatility of glass is an important advantage to its use in medicine [30] since its properties can be varied over a wide range and tailored to a particular application by simple changes in its chemical composition. Glass microspheres can range from bio-inert to bioactive, from almost completely insoluble to biodegradable in the body. Similarly, the radiation emitted by glass microspheres can be tailored for a particular organ by using a glass that contains one or more neutron activatable elements (e.g. REs) that emit the optimum radiation for the organ being treated. The manufacturing of glass microspheres is much easier when the microsphere is made radioactive by neutron activation of the desired radioisotope as the last step in the manufacturing process.

One major and unique advantage of using radioactive glass microspheres to irradiate diseased organs in *situ* is that very large doses of beta radiation can be safely delivered to a patient in this way, with minimum damage to healthy tissue. Patients with primary liver cancer are routinely receiving a calculated whole liver dose of up to 150 Gy, with no major side effects. Some patients have received multiple doses of 150 Gy and there are a few documented cases where localized regions of HCC tumors have received from 1000 to 8000 Gy with no adverse side effects. One reason why such large doses can be used is that the radioactive glass microspheres are preferentially deposited in the tumor as opposed to the healthy tissue at ratios from 10 : 1 to 40 : 1. Another important factor is that the shorter-range, more localized beta radiation can be used instead of the higher energy radiation used in external beam methods.

Based on animal experiments and the results from treating patients with inoperable liver cancer (HCC), there is reason to believe that radioactive glass microspheres can also be used to destroy malignant tumors in other organs such as the kidney, brain, breast, prostate, and pancreas. It is conceivable that glass microspheres can be engineered to function as long-term drug delivery vehicles in combination with *in situ* irradiation therapy. In other medical applications, it might be desirable to implant fibers of a radioactive glass in a particular geometric pattern in the organ so that the organ is irradiated from a series of line sources instead of from point sources, microspheres.

In summary, ^{90}YAS glass microspheres offer a safe and reliable way of delivering unusually large doses of beta radiation to malignant tumors in the liver with minimum side effects. The procedure of *in situ* irradiation has extended the life expectancy of patients with inoperable liver cancer (HCC), with some patients surviving more than five years since treatment with the radioactive glass microspheres.

REFERENCES

[1] Ehrhardt, G.J. and Day, D.E. (1987) Therapeutic use of yttrium-90. *Nuclear Medicine and Biology International Journal of Radiation Applications and Instrumentation Part B*, **14**, 233–242.

[2] Houle, S., Yip, T.K., Shepherd, F.A. *et al.* (1989) Hepatocellular carcinoma: pilot trial of treatment with Y-90 microspheres. *Radiology*, **172**, 857–860.

[3] Shepherd, F.A., Rotstein, L.E., Yip, T.K. *et al.* (1992) A phase I dose escalation trial of yttrium-90 microspheres in the treatment of primary hepatocellular carcinoma. *Cancer*, **70**, 2250–2254.

[4] White, J.E. and Day, D.E. (1994) Rare earth aluminosilicate glasses for in vivo radiation delivery. *Key Engineering Materials*, **94–95**, 181–208.

[5] Day, D.E. and Day, T.E. (1993) Radiotherapy glasses. In *An Introduction to Bioceramics*, Advanced Series in Ceramics (eds L.L. Hench and J. Wilson). Singapore: World Scientific, pp. 305–317.

[6] Erbe, E.M. and Day, D.E. (1993) Chemical durability of Y_2O_3–Al_2O_3–SiO_2 glasses for the in-vivo delivery of beta radiation. *Journal of Biomedical Materials Research*, **27**, 1301–1308.

[7] McIntyre, D.S. and Day, D.E. (1998) Ho_2O_3–Al_2O_3–SiO_2 glasses for in-vivo radiotherapy. *Physics and Chemistry of Glasses*, **39**, 29–35..

[8] Shelby, J.E. (1994) Rare elements in glasses. *Key Engineering Materials*, **94–95**, 1–420.

[9] Conzone, S.D., Brown, R.F., Day, D.E., and Ehrhardt, G.J. (2002) In vitro and in vivo dissolution behavior of a dysprosium lithium borate glass designed for the radiation synovectomy treatment of rheumatoid arthritis. *Journal of Biomedical Materials Research*, **60**, 260–268.

[10] Day, D.E., White, J.E., Brown, R.F., and McMenamin, K.D. (2003) Transformation of borate glasses into biologically useful materials. *Glass Technology*, **44**, 75–81.

[11] Day, D.E. and White, J.E. (2002) Biodegradable glass compositions & methods for radiation therapy. US Patent 6,379,648.

[12] Day, D.E. and Conzone, S.A. (2002) Method for preparing porous shells or gels from glass particles. US Patent 6,358,531.

[13] Hafeli, U.O., Casillas, S., Dietz, D.W. *et al.* (1999) Hepatic tumor radioembolization in a rat model using radioactive rhenium (^{186}Re/^{188}Re) glass microspheres. *International Journal of Radiation Oncology Biology Physics*, **44**, 189–199.

[14] Conzone, S.D., Hall, M.M., Day, D.E., and Brown, R.F. (2004) Biodegradable radiation delivery system utilizing glass microspheres and ethylenediaminetetraacetate chelation therapy. *Journal of Biomedical Materials*, **70A**, 256–264.

[15] Wollner, I., Knutsen, C., Smith, P. *et al.* (1988) Effects of hepatic arterial yttrium 90 glass microspheres in dogs. *Cancer*, **61**, 1336–1343.

[16] Salem, R., Thurston, K.G., Carr, B.I. *et al.* (2002) Yttrium-90 microspheres: radiation therapy for unresectable liver cancer. *Journal of Vascular and Interventional Radiology*, **13**, S223–S229.

[17] Anderson, J.H., Goldberg, J.A., Bessent, R.G. *et al.* (1992) Glass yttrium-90 microspheres for patients with colorectal liver metastases. *Radiotherapy and Oncology*, **25**, 137–139.

[18] Kennedy, A.S., Nutting, C., Coldwell, D. *et al.* (2004) Pathologic response and microdosimetry of ^{90}Y microspheres in man: review of four explanted whole livers. *International Journal of Radiation Oncology Biology Physics*, **60**, 1552–1563.

[19] Brown, R.F., Lindesmith, L.C., and Day, D.E. (1991) ^{166}Holmium-containing glass for internal radio therapy of tumors. *Nuclear Medicine and Biology*, **18**, 783–790.

[20] Qian, X., Zhou, N., Huang, W. *et al.* (1998) Preparation of radio-therapeutical glass microspheres for curing malignant tumor: V. The study of radiant characteristics and animal experiments. *Digest of Chinese Academic Journals – Review of Science and Technology*, **4**, 1012–1014.

[21] Andrews, J.C., Walker, S.C., Ackermann, R.J. *et al.* (1994) Hepatic radioembolization with yttrium-90 containing glass microspheres: preliminary results and clinical follow-up. *Journal of Nuclear Medicine*, **35**, 1637–1644.

[22] Ariel, I.M. (1965) Treatment of inoperable primary pancreatic and liver cancer by the intra-arterial administration of radioactive isotopes (Y^{90} radiating microspheres). *Annals of Surgery*, **162**, 267–278.

[23] Kim, Y., LaFave, J.W., and MacLean, L.D. (1962) The use of radiating microspheres in the treatment of experimental and human malignancy. *Surgery*, **52**, 221–231.

[24] Nolan, T.R. and Grady, E.D. (1969) Intravascular particulate radioisotope therapy clinical observations of 76 patients with advanced cancer treated with 90-yttrium particles. *The American Surgeon*, **35**, 181–189.

[25] Ehrhardt, G.J., Curtis, R.L., Latimer, J.C. *et al.* (2000) Investigation of pre-operative sterilization of kidney cancers using intra-arterial samarium-153 microspheres. *University of Missouri Research Reactor*, **845**, 1024.

[26] Spooren, P.F.M.J., Rasker, J.J., and Arens, R.P.J.H. (1985) Synovectomy of the knee with ^{90}Y. *European Journal of Nuclear Medicine*, **10**, 441–445.

[27] Day, D.E. (1995) Reactions of bioactive borate glasses with physiological liquids. *Glass Researcher*, **12**, 21–22.

[28] White, J.E., Day, D.E., Brown, R.E., and Ehrhardt, G.J. (2010) Biodegradable rare earth lithium aluminoborate glasses for brachytherapy use. In *Advances in Bioceramics and Porous Ceramics III* (eds R. Narayan, P. Colombo, S. Mathur, and T. Ohji), Ceramic Engineering and Science Proceedings, vol. 31. Hoboken, NJ: John Wiley & Sons, Inc., pp. 3–18.

[29] Shortkroff, S. (2000) The influence of radionuclides on synovitis and its assessment by MRI. PhD thesis. University of Bristol, Bristol, UK.

[30] Hench, L.L., Day, D.E., Holand, W., and Rheinburger, V.M. (2010) Glass and medicine. *International Journal of Applied Glass Science*, **1**, 104–117.

Index

References to figures are given in italic type. References to tables are given in bold type.

Bio-Glasses: An Introduction, First Edition. Edited by Julian R. Jones and Alexis G. Clare.
© 2012 John Wiley & Sons, Ltd. Published 2012 by John Wiley & Sons, Ltd.